A More PERFECT HEAVEN

Also by Dava Sobel

Longitude

The Illustrated Longitude
(with William J. H. Andrewes)

Galileo's Daughter

The Planets

Letters to Father (translated and annotated)

GLOBI SUBLUNARIS CUM LUNA ET QUATUOR
CIRCULUS
MARS
SATURNI
MARTIS

A More PERFECT HEAVEN

How Copernicus *Revolutionized the Cosmos*

DAVA SOBEL

BLOOMSBURY
LONDON • BERLIN • NEW YORK • SYDNEY

First published in Great Britain 2011

For illustration credits, see page 263

Bloomsbury Publishing Plc
50 Bedford Square
London WC1B 3DP

www.bloomsbury.com

Bloomsbury Publishing, London, New York and Berlin

A CIP catalogue record for this book
is available from the British Library

Hardback ISBN 978 1 4088 1800 8
Trade paperback ISBN 978 1 4088 1983 8

10 9 8 7 6 5 4 3 2

Printed in Great Britain by Clays Ltd, St Ives plc

To my fair nieces,

AMANDA SOBEL

and

CHIARA PEACOCK,

with love in the Copernican

tradition of nepotism.

Contents

Part Three · *Aftermath*

CENTRAL EUROPE
in the 16th Century
Uppsala
Stockholm
SWEDEN
North Sea
DENMARK
Hven
Copenhagen
Baltic Sea
LITHUANIA
Königsberg
DUCAL PRUSSIA
Danzig
Frauenburg
Allenstein
ROYAL PRUSSIA
Torun
Bremen
Elbe River
BRANDENBURG
Berlin
SPANISH NETHERLANDS
Weser River
Wittenberg
Warsaw
POLAND
Rhine River
HOLY
SAXONY
Leipzig
SILESIA
Oder River
Vistula R.
Saalfeld
Frankfort
Prague
Krakow
ROMAN
BOHEMIA
Heidelberg
Nuremberg
HAPSBURG
CARPATHIAN MTNS.
LORRAINE
Tübingen
Danube River
MORAVIA
ROYAL HUNGARY
BAVARIA
EMPIRE
Cassovia
EMPIRE
Munich
Vienna
Basel
Lindau
AUSTRIA
Budapest
Zurich
Feldkirch
SWITZERLAND
ALPS
HUNGARY
VENICE
Milan
SAVOY
MILAN
Padua
Venice
Danube River
Po River
Ferrara
APENNINES
Bologna
OTTOMAN
EMPIRE
Adriatic Sea
Florence
TUSCANY
PAPAL STATES
0 Miles 100 200
0 Kilometers 200
© 2011 Jeffrey L. Ward
Rome

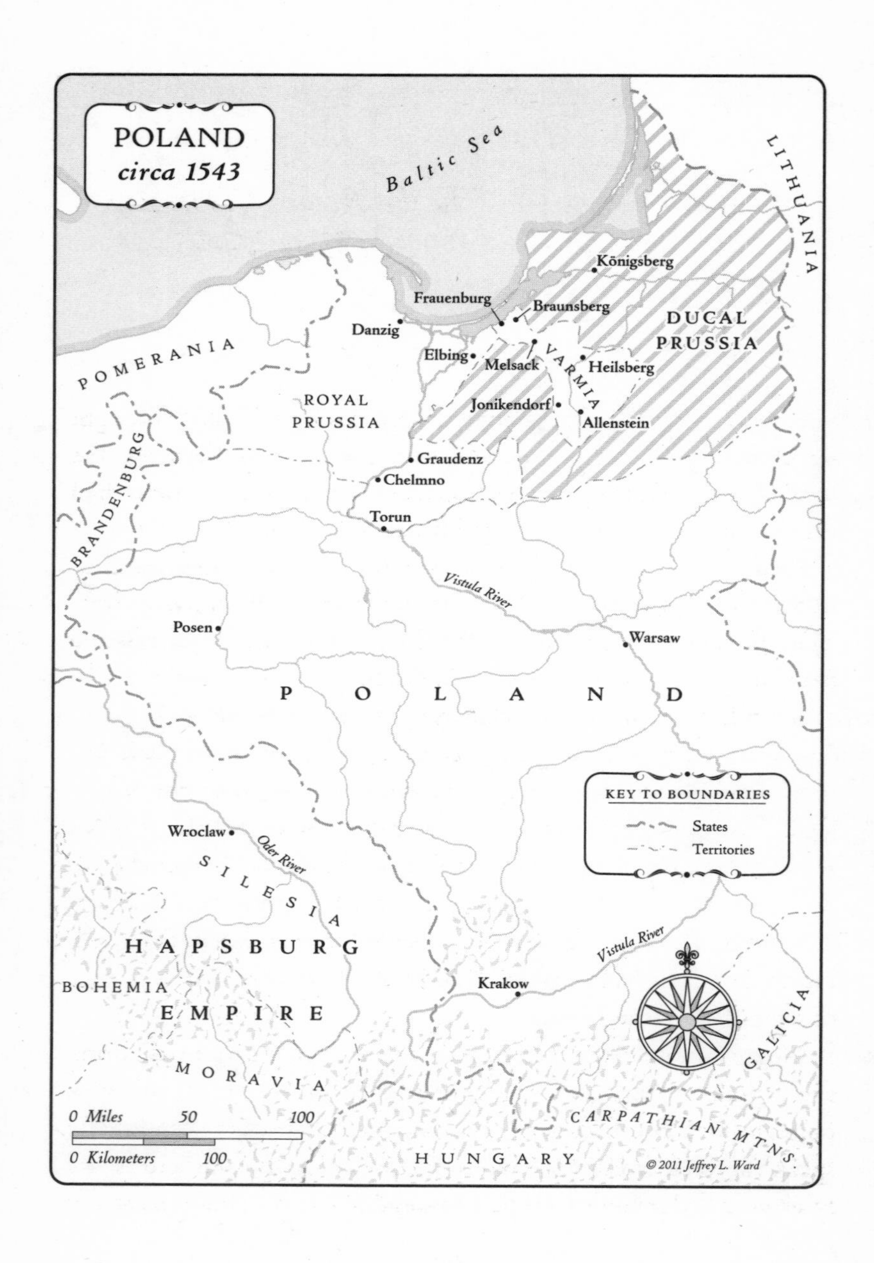
POLAND
circa 1543
Baltic Sea
LITHUANIA
Königsberg
Frauenburg
Braunsberg
DUCAL
PRUSSIA
Danzig
Elbing
Melsack
VARMIA
Heilsberg
POMERANIA
ROYAL
PRUSSIA
Jonikendorf
Allenstein
BRANDENBURG
Graudenz
Chelmno
Torun
Vistula River
Posen
Warsaw
P O L A N D
KEY TO BOUNDARIES
States
Territories
Wroclaw
Oder River
SILESIA
HAPSBURG
Vistula River
BOHEMIA
Krakow
EMPIRE
GALICIA
MORAVIA
0 Miles 50 100
CARPATHIAN MTNS.
0 Kilometers 100
HUNGARY
© 2011 Jeffrey L. Ward

"To the Reader, Concerning . . . This Work"

SINCE 1973, when the five hundredth anniversary of his birth brought his unique story to my attention, I have wanted to dramatize the unlikely meeting between Nicolaus Copernicus and the uninvited visitor who convinced him to publish his crazy idea.

Around the year 1510, near the age of forty, Copernicus reenvisioned the cosmos with the Sun, rather than the Earth, at its hub. Then he concealed the theory for thirty years, fearful of ridicule from his mathematician peers. But when his unexpected guest, called Rheticus, made the dangerous, several-hundred-mile journey to northern Poland in 1539, eager to learn the novel planetary order from its source, the aging Copernicus agreed to end his silence. The youth stayed on for two years, despite laws barring his presence, as a Lutheran, from Copernicus's Catholic diocese during this contentious phase of the Protestant Reformation. Rheticus helped his mentor prepare the long-neglected manuscript for publication, and later hand-carried it to Nuremberg, to the best printer of scientific texts in Europe.

No one knows what Rheticus said to change Copernicus's mind about going public. Their dialogue in the two-act play that begins on page 81 is my invention, although the characters occasionally speak the very words they wrote themselves in various letters and treatises. I had intended the play to stand on its own, but I thank my perceptive editor, George Gibson, for urging me to plant it in the broad context of history by surrounding the imagined scenes with a fully documented factual narrative that tells Copernicus's life story and traces the impact of his seminal book, *On the Revolutions of the Heavenly Spheres*, to the present day.

Part One
Prelude

❧ ❧

Bless the Lord, O my soul.
Who layeth the beams of his chambers in the waters: who maketh the clouds his chariot: who walketh upon the wings of the wind.
Who laid the foundations of the Earth, that it should not be removed for ever.

—Psalm 104:1, 3, 5

The great merit of Copernicus, and the basis of his claim to the discovery in question, is that he was not satisfied with a mere statement of his views, but devoted a large part of the labor of a life to their demonstration, and thus placed them in such a light as to render their ultimate acceptance inevitable.

—From *Popular Astronomy* (1878),
by Simon Newcomb, founding president of
the American Astronomical Society

CHAPTER 1

Moral, Rustic, and Amorous Epistles

The cricket is a musical being. At the break of dawn it starts to sing. But much louder and more vociferous, according to its nature, is it heard at the noon hour, because intoxicated by the Sun's rays. As the songster chirps, then, it turns the tree into a platform and the field into a theater, performing a concert for the wayfarers.

—FROM *Letters of Theophylactus Simocatta*,
THE FIRST PUBLISHED WORK BY COPERNICUS, 1509

ICOLAUS COPERNICUS, the man credited with turning our perception of the cosmos inside out, was born in the city of Torun, part of "Old Prussia" in the Kingdom of Poland, at 4:48 on Friday afternoon, the nineteenth of February, 1473. His horoscope for that auspicious moment (preserved in the Bavarian State Library in Munich) shows the Sun at 11° of Pisces in the sixth house, while Jupiter and the Moon are "conjunct," or practically on top of one another, at 4° and 5°, respectively, of Sagittarius, in the third house. Whatever clues to character or destiny such data may contain, this particular natal chart is an after-the-fact construct, created at the end of the astronomer's life and not the beginning of it (with the time of birth calculated, as opposed to copied from a birth certificate). At the time his horoscope was cast, Copernicus's contemporaries already knew he had fathered an alternate universe—that he had defied common sense and received wisdom to place the Sun at the center of the heavens, then set the Earth in motion around it.

Nearing seventy, Copernicus had little cause to recall the exact date of his birth, let alone the hour of it down to the precision of minutes. Nor had he ever expressed the slightest faith in any astrological prognostications. His companion at the time, however, a professed devotee of the "juridical art," apparently pressed Copernicus for biographical details to see how his stars aligned.

The horoscope's symbols and triangular compartments position the Sun, Moon, and planets above or below the horizon, along the zodiac—the ring of constellations through which they appear to wander. The numerical notations describe more precisely where they lie at the moment, with respect to the twelve signs and also twelve so-called houses governing realms of life experience. Although the diagram invites interpretation, no accompanying conjecture has survived alongside it. One modern astrologer, invited to consider Copernicus's case, used computer software to draw a new configuration in the shape of a wheel, and added solar-system bodies unknown in his time. Uranus and Neptune thus crept into the third house beside the Moon and Jupiter, while Pluto, a dark force, manifested itself opposite the Sun, at 16° of Virgo in the first house. The Pluto-Sun opposition drew a gasp from the astrologer, who declared it the hallmark of a born revolutionary.

The bold plan for astronomical reform that Copernicus conceived and then nurtured over decades in his spare time struck him as the blueprint for the "marvelous symmetry of the universe." Even so, he proceeded cautiously, first leaking the idea to a few fellow mathematicians, never trying to proselytize. All the while real and bloody revolutions—the Protestant Reformation, the Peasant Rebellion, warfare with the Teutonic Knights and the Ottoman Turks—churned around him. He held off publishing his theory for so long that when his great book, *On the Revolutions of the Heavenly Spheres*, finally appeared in print, its author breathed his last. He never heard any of the criticism, or acclaim, that attended *On the Revolutions*. Decades after his death, when the first telescopic dis-

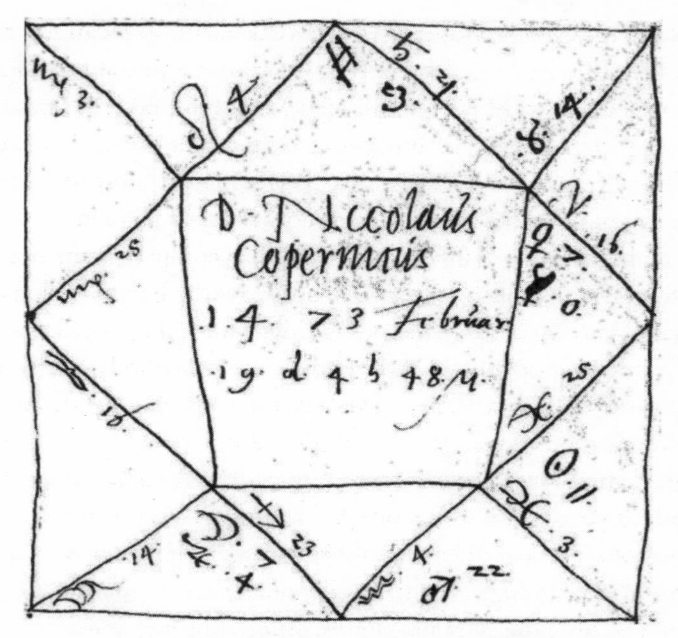

HOROSCOPE FOR NICOLAUS COPERNICUS
Astronomers and astrologers in Copernicus's time shared the same pool of information about the positions of the heavenly bodies against the backdrop of the stars. Until the invention of the telescope in the seventeenth century, position finding and position predicting constituted the entirety of planetary science—and the basis for casting horoscopes.

coveries lent credence to his intuitions, the Holy Office of the Inquisition condemned his efforts. In 1616, *On the Revolutions* was listed on the *Index of Prohibited Books*, where it remained for more than two hundred years. The philosophical conflict and change in perception that his ideas engendered are sometimes referred to as the Copernican Revolution.

He was christened for his father—Mikolaj in Polish, Niklas in German, his native tongue. Later, as a scholar, he Latinized his name, but he grew up Niklas Koppernigk, the second son and youngest child of a merchant family from the copper-mining regions of Silesia. Their ancestral village of Koperniki could have taken its name from the Slavic word for the dill plant, *koper*, or from the Old German term for the metal mined there, *kopper*—or maybe it commemorated both those products of its hillsides. In any case, the roots of Koperniki's etymology lay long buried by the time its younger generations began leaving home to seek new fortunes in the towns and cities. An armorer named Mikolaj Kopernik appeared in the city chronicles of Krakow in 1375, followed by mention of the mason Niclos Kopernik in 1396 and the rope maker Mikolaj Kopernik in 1439, all bearing the names of their forefathers' homeland and its popular patron saint.

Around the year 1456, the alderman Mikolaj Koppernigk, who traded in Hungarian copper, moved north from Krakow to Torun, where he married Barbara Watzenrode. They lived on narrow St. Anne's Lane, later renamed Copernicus Street, and raised four children in a tall brick house that is now a museum to the memory of their famous son. From the double front doors under the house's pointed arch, their two boys, Andrei and Niklas, could walk to classes at the parish school of St. John's Church, or down to the family warehouse near the wide river, the Vistula, that coursed from Krakow past Warsaw through Torun, carrying the flow of commerce to Danzig on the Baltic Sea.

Soon after the boy Niklas reached ten years of age, the elder Niklas died. His bereft sons and daughters and his widow, Barbara Koppernigk, turned for succor to her brother, Lukasz Watzenrode, a minor cleric, or "canon," in a nearby diocese. Or perhaps Barbara, whose date of death is not recorded, had predeceased her husband, leaving her brood true orphans. Either way, the children came under their uncle's care. Canon Watzenrode arranged a marriage contract for his niece Katyryna with Bartel Gertner of Krakow and consigned his niece Barbara to the Cistercian convent at Kulm. His young nephews he supported at school, first in Torun and later in

Kulm or Wloclawek, until they were ready to attend his alma mater, the Jagiellonian University in Krakow. By then Uncle Lukasz had risen from a mediocre position in the Catholic hierarchy to become Bishop of Varmia.

A page of Gothic script in the archives of the Collegium Maius at the Jagiellonian University attests that Nicolaus Copernicus, age eighteen, paid his tuition fees in full for the fall of 1491. He studied logic, poetry, rhetoric, natural philosophy, and mathematical astronomy. According to the courses in his curriculum, his father's copper and other common substances could not be considered elements in the modern sense of the periodic table. Rather, they comprised some combination of the four classic elements: earth, water, air, and fire. The heavens, in contrast, consisted entirely of a fifth essence, called ether, which differed from the other four by virtue of being inviolate and everlasting. Ordinary objects on Earth moved more or less along straight paths, whether seeking their natural places in the world order or being compelled by outside agents. Heavenly bodies, however, lay cocooned in celestial spheres that spun in eternal perfect circles.

The motions of the planets captured Copernicus's interest from the start of his university studies. At college he purchased two sets of tables for calculating their positions and had these bound together, adding sixteen blank pages where he copied parts of a third table and wrote miscellaneous notes. (This custom volume and other remnants of his personal library, seized as spoils of the Thirty Years' War, now belong to the University of Uppsala, Sweden.) Copernicus more than once explained his attraction to astronomy in terms of beauty, asking rhetorically, "What could be more beautiful than the heavens, which contain all beautiful things?" He also cited the "unbelievable pleasure of mind" he derived from contemplating "things established in the finest order and directed by divine ruling."

"Among the many various literary and artistic pursuits upon which the natural talents of man are nourished," he wrote, "I think the ones above all to be embraced and pursued with the most loving care

ARISTOTLE'S UNIVERSE

As Copernicus learned in school, the world around him consisted of the four elements: earth, water, air, and fire. Far removed from these ordinary substances, the Moon and other celestial bodies consisted of a fifth essence, immune to change or destruction. In the perfect heavens, bodies moved with uniform circular motion.

concern the most beautiful and worthy objects, most deserving to be known. This is the nature of the discipline that deals with the god-like circular movements of the world and the course of the stars."

The portrait of him now hanging in Torun's town hall cuts a youthful, handsome figure. Based on a purported self-portrait that disappeared long ago, it shows Copernicus dressed in a red jerkin, with glints painted into his dark eyes and dark hair. (The light in each brown iris reflects, on close inspection, the tall Gothic windows of the rooms he frequented.) He had a long nose, a manly shadow above his full lips, and a faint scar extending from the corner of his left eye up into the eyebrow. This mark encouraged archaeologists in 2005, who picked out his skull among the litter of remains under the church where he had lain buried. A double dent above the skull's right eye socket—not the left one—seemed to affirm their identification, since every portraitist sees himself as his mirror's image.

In September 1496, again at his uncle's command, Copernicus traveled to Italy to study canon law, concerning the rights and duties of the clergy, at the University of Bologna. Only one year into this enterprise, he became a canon himself. The death of one of the sixteen Varmia canons created a vacancy, and Bishop Watzenrode used his connections to win Copernicus the office in absentia. As the fourteenth canon of the Cathedral Chapter—effectively a trustee in the rich and powerful governing body of the Varmia diocese—Copernicus could now collect an income independent of his allowance.

He lodged in Bologna with the local astronomy professor, Domenico Maria Novara, whom he assisted in nightly observations. Together they watched the Moon pass in front of the bright star Aldebaran (the eye of Taurus the Bull) on March 9, 1497, and Copernicus recorded in his notes how the star hid "between the horns of the moon at the end of the fifth hour of the night."

At the conclusion of his law studies, he visited Rome in the summer of 1500 for the jubilee year celebrations. He and other pilgrims tripled the population of the Holy City, where a crowd of two hundred thousand knelt to receive the Easter Sunday blessing of Pope Alexander VI. Still in Rome on November 6, Copernicus observed and recorded a partial lunar eclipse. He also lectured in Rome about mathematics to students and experts alike. But his future with the Church had already been decided. July 27, 1501, found him at a meeting of the Cathedral Chapter in Varmia, along with his older brother, Andreas, who had also attained a canonry there, courtesy

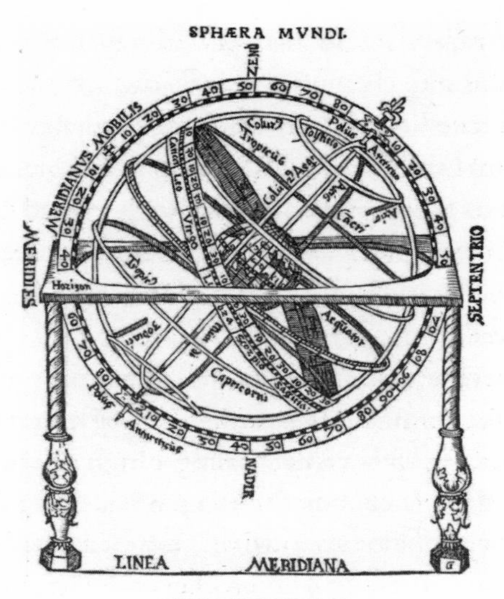

THE ZODIAC

The Earth-centered universe that Copernicus inherited is pictured in this frontispiece from one of his favorite books, the Epitome of Ptolemy's Almagest, *by Regiomontanus. He and other astronomers measured the motions of the "wandering" stars—the planets, the Sun, and the Moon—through the band of "fixed" stars called the zodiac. The Sun took about one month to progress through each sign, completing the circuit from the ram, Aries, to the fishes, Pisces, in a year. Since the actual constellations vary considerably in size, astronomers arbitrarily assigned the same one twelfth of a circle, or 30°, to each zodiac sign.*

of Uncle Lukasz. Both young men requested leave to return to Italy for further education and received the chapter's blessing. They set out almost immediately for Padua, where Copernicus studied medicine in preparation for a career as "healing physician" to the bishop and canons of Varmia.

In his novel *Doctor Copernicus*, John Banville imagines the brothers equipping themselves for their journey "with two stout staffs, good heavy jackets lined with sheepskin against the Alpine cold, a tinderbox, a compass, four pounds of sailor's biscuit and a keg of salt pork."

This and other rich descriptions—one of which pictures "Nicolas" sewing gold coins into the lining of his cloak for safekeeping—leap the gaps in the true life story. Historians have pieced that together from his few published works and the scattered archives where he left his name. His lifetime of correspondence comes down today to just seventeen surviving signed letters. (Of these, three concern the woman who lived with him as cook and housekeeper, and probably concubine as well.)

"The inns were terrible, crawling with lice and rogues and poxed whores," Banville continues the brothers' travel narrative. "And then one rainy evening as they were crossing a high plateau under a sulphurous lowering sky a band of horsemen wheeled down on them, yelling. They were unlovely ruffians, tattered and lean, deserters from some distant war. . . . The brothers watched in silence their mule being driven off. Nicolas's suspiciously weighty cloak was ripped asunder, and the hoard of coins spilled out." It could all have happened, just that way.

As a medical student at the University of Padua, Copernicus learned therapeutic techniques, such as bloodletting with leeches, aimed at balancing the four bodily humors: blood, phlegm, black bile, and yellow bile. All manifestations of health or disease stemmed from an excess or deficiency of one or more of these fluids. Even gray hair was caused by "corrupt humors" and could be postponed with the proper prescription. Copernicus also watched anatomical dissections, studied surgical procedures, and took instruction in the application of astrology to diagnosis and treatment. His textbooks, which were still with him at his death and mentioned in his will, included the 1485 edition of *Breviarium practicae* by Arnaldus of Villanova, a thirteenth-century physician and alchemist.

"To produce sleep so profound that the patient may be cut and will feel nothing, as though he were dead," Arnaldus advised, "take of opium, mandragora bark, and henbane root equal parts, pound them together and mix with water. When you want to sew or cut a man, dip a rag in this and put it to his forehead and nostrils. He will soon sleep so deeply that you may do what you will. To wake him up, dip the rag in strong vinegar."

Copernicus cut short his medical studies after two of the required three years. Having never been graduated from any of the universities he attended, he traveled to Ferrara in May of 1503, sat for the exam in canon law, and took his doctoral degree. Some Copernicus scholars think he did this to avoid the hoopla of commencement rituals in the university courtyard at Padua, "Il Bo," not to mention the cost of the fees paid to examiners and the dinner party that a new graduate was expected to throw. From Ferrara he returned to Poland—to Varmia—for good.

The cathedral of Varmia stood, as it still stands today, on a hilltop overlooking the Vistula Bay. The great brick church rises in Gothic turrets and spires from a stone foundation laid in the fourteenth century. A few small buildings, a bell tower, and a covered well huddle around the church, surrounded in turn by high fortified walls, crowned with crenellations and arrow loops. The moat and barbican are gone, but the gateways retain the thick, grudging wooden doors and medieval grates that even now can fall with fatal weight.

The presence of the cathedral dedicated to the Virgin Mary gave the name Frauenburg, or "the city of Our Lady," to the adjacent community. Frauenburg (known today as Frombork), was one of several cities within the diocese of Varmia. The imposing bishop's palace, where Doctor Copernicus first went to live and work for his uncle, lay fifty miles away, in Heilsberg (now Lidzbark Warmiński). The fifty-mile remove seems extremely inconvenient, given that it took days to travel such a distance at the pace of available transport, but Bishop Watzenrode was only occasionally required to appear at the cathedral. On January 11, 1510, for example, he arrived there leading an official procession, having carried the sacred relic said to be St. George's head all the way from Heilsberg.

As much a prince as a prelate, the Bishop of Varmia governed a province of more than four thousand square miles (most of which belonged to him personally) with tens of thousands of inhabitants. He reported directly to the King of Poland. Indeed, Watzenrode served as trusted counselor to three successive kings over the course of his episcopate, sharing with them his dreams of Polish glory and his hatred for the white-cloaked Knights of the Teutonic Order, whose lands engulfed Varmia. Although the military-religious order had been founded in the Holy Land by Crusaders late in the twelfth century, it removed after the fall of Acre to Old Prussia, where it grew dissolute and dangerous. Often the knights thundered out of their castle at Königsberg to raid the towns of Varmia—even attacking Frauenburg and its cathedral fortress.

Bishop Watzenrode had fathered an illegitimate son in Torun, but he regarded his talented younger nephew as his heir apparent. Having nurtured Copernicus through Church ranks, he now positioned him as episcopal physician and personal secretary, poised on the brink of limitless advancement. Yet the youth seemed not nearly hungry enough. His mind strayed from the lanes of power, as suggested by the notes Copernicus kept from his years in the bishop's employ. These describe the positions of Mars, Jupiter, and Saturn during their Great Conjunction in the sign of Cancer in 1504, and the lunar eclipse that occurred on June 2, 1509.

French polymath Pierre Gassendi, who wrote the first extant biography of Copernicus in 1654, more than a century after the astronomer's death, said he treated the illnesses of the poor without charging them any fee. While it is easy and tempting to presume the goodness of his heart, the peasants of Varmia probably had no money to pay for his services, nor he any need of their pennies. In addition to the income from his canonry, Copernicus received a second livelihood from a sinecure at the Church of the Holy Cross in Wroclaw, which he retained for thirty-five years. Also, the Cathedral Chapter of Varmia paid him an annual bonus for tending to the bishop's medical complaints. Records show that when Bishop Watzenrode took sick in 1507, his nephew successfully restored him to health.

Copernicus made a public display of gratitude to his uncle by dedicating his first published work to him, hailing Watzenrode as "O right reverend ruler and father of our country." The text thus offered was not the great Copernican theory but a translation, from Greek into Latin, of a collection of letters by a seventh-century moralist from Constantinople. Copernicus found the eighty-five moral, rustic, and amorous letters of Theophylactus Simocatta in the chapter library, in a volume called *Epistolographers*. The missives read more like fables and teachings than communiqués, but he liked them, he said, because "Theophylactus so interspersed the gay with the serious, and the playful with the austere, that every reader may pluck what pleases him most in these letters, like an assortment of flowers in a garden."

One of the letters dealt specifically with an uncle's duty to a nephew: "Among mares there is a rule, and it seems to me quite wise. Indeed I praise their profound kindliness. But what is this rule? If they see a foal lacks a teat and the mother is far away, any one of them nurses the foal. For they do not forget their own species and, with a single purpose and no ill will, they do their nursing as though having to do with their own true descendant. . . .

"Now I shall apply this discourse to you. You scorn your brother's son as he roams from door to door, clad in most wretched rags. Your feelings are less sensible than the brutes'. You feed others' hounds, for that is what I would quite properly call the flatterers around you. For they appear to be completely loyal as long as they are stuffed full of your food, you wretch! Yet they constantly bark at you even while they are still belching out the booze they just drank. For, flatterers constitute a breed that is mindful of harm and most forgetful of favors. Therefore . . . take care of your nephew at last. If you do not, you will have your conscience as your implacable foe, sharpening his sword with Nature's tears."

Fortunately for Copernicus, his own Uncle Lukasz had needed no such admonition to extend a generous hand.

Anxious about the reception of the erotic letters, Copernicus claimed to have cleaned them up for the bishop's sake: "Just as physicians usually modify the bitterness of drugs by sweetening them to make them more palatable to patients," he wrote in the dedication, "so these love letters have in like manner been rectified." Even so, they make mention of lust, carnal desires, irrational passion, prostitution, infidelity, abortion, and infanticide.

A friend of Copernicus, Wawrzyniec Korwin (pen name Laurentius Corvinus), took the manuscript of the little book to Krakow for printing in 1509. At that date, no press had yet been established anywhere in Varmia, or even in Torun. Korwin also wrote an introductory poem for the work. His verses provided a character assessment of the bishop—"conspicuous for his piety" and "revered for his grave demeanor"—suggesting that Watzenrode may have dispensed generosity without much personal warmth. As for "the scholar who translates this work," Korwin knew him to be engaged in loftier pursuits: "He discusses the swift course of the Moon and the alternating movements of its brother as well as the stars together with the wandering planets—the Almighty's marvelous creation—and he knows how to seek out the hidden causes of phenomena by the aid of wonderful principles."

Copernicus was already reconceiving the order of the heavenly spheres. In fact, the whole exercise of teaching himself Greek—and practicing his proficiency on the moral, rustic, and amorous letters of Theophylactus Simocatta—seems to have been a prerequisite for studying the works of Greek astronomers and consulting the ancient Greek/Egyptian calendar, in order to date correctly their observations from antiquity.

In mid-1510, Copernicus somehow communicated to the reigning Bishop of Varmia that he did not aspire to become the future one, for he moved out of the palace. After relocating near the cathedral in Frauenburg, he no longer accompanied his uncle on diplomatic missions—not even to Krakow in February 1512 for the wedding of King Sigismund and the coronation of his new queen, the young Hungarian noblewoman Barbara Zapolya. Bishop Watzenrode doubtless rued his nephew's absence from these festivities, especially on the return journey, when he fell ill with a fever. He stopped at Torun, hoping to recuperate there before continuing on to Heilsberg, but his condition only grew worse. He died three days later, on March 29, at sixty-four years of age.

The last of Theophylactus's letters had touched on death and its lessons for the living. "Stroll through the tombstones," it counseled those weighed down by their own sorrows. "You will behold man's greatest joys as in the end they take on the lightness of dust."

Chapter 2
The Brief Sketch

The center of the earth is not the center of the universe, but only the center towards which heavy things move and the center of the lunar sphere.

—From the *Commentariolus*, or *Brief Sketch*,
by Copernicus, ca. 1510

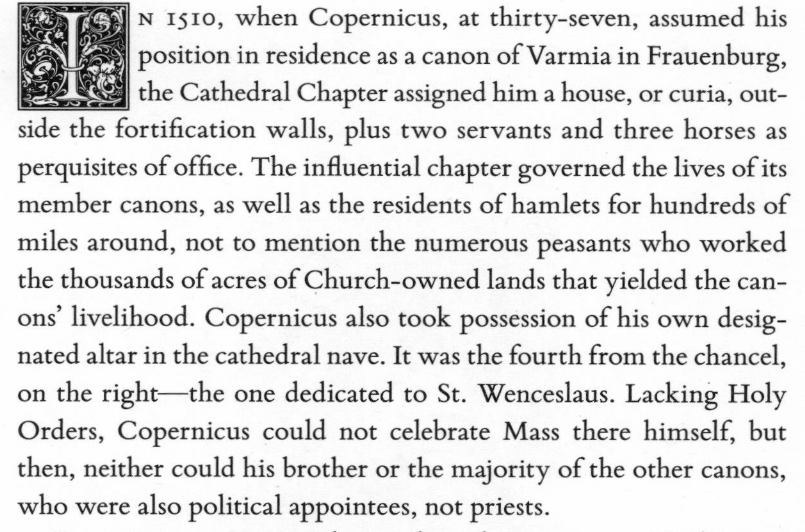

IN 1510, when Copernicus, at thirty-seven, assumed his position in residence as a canon of Varmia in Frauenburg, the Cathedral Chapter assigned him a house, or curia, outside the fortification walls, plus two servants and three horses as perquisites of office. The influential chapter governed the lives of its member canons, as well as the residents of hamlets for hundreds of miles around, not to mention the numerous peasants who worked the thousands of acres of Church-owned lands that yielded the canons' livelihood. Copernicus also took possession of his own designated altar in the cathedral nave. It was the fourth from the chancel, on the right—the one dedicated to St. Wenceslaus. Lacking Holy Orders, Copernicus could not celebrate Mass there himself, but then, neither could his brother or the majority of the other canons, who were also political appointees, not priests.

Copernicus encountered an ordained exception in Tiedemann Giese, a fellow canon seven years younger than he. Giese came from a well-known family in Danzig, where he had presided at the Church of Peter and Paul. He shared with Copernicus an abiding interest in astronomy, perhaps acquired when they befriended each other. Giese was almost certainly the first to hear Copernicus confess his secret knowledge of the cosmos. One imagines the priest's initial reaction to those unorthodox ideas as skeptical at best, but in time he concurred, even encouraged Copernicus, and urged him to disseminate his theory.

By 1510, Copernicus had leapt to his Sun-centered conclusion via intuition and mathematics. No astronomical observations were required. He wrote out a short overview of his new heavenly arrangement, also probably in 1510, and sent it off to at least one correspondent beyond Varmia. That person in turn copied the document for further circulation, and presumably the new recipients did, too, because by May of 1514, when the Krakow physician and medical professor called Matthew of Miechow inventoried his private library, it contained "A manuscript of six leaves containing a Theorica [astronomy essay] in which the author asserts that the Earth moves while the Sun stands still."

Copernicus had no idea that Aristarchus of Samos had proposed much the same thing in the third century B.C. The only work by

Tiedemann Giese,
canon of Varmia.

Aristarchus known to Copernicus—a treatise called *On the Sizes and Distances of the Sun and Moon*—made no mention of a heliocentric plan. Copernicus stood alone, for the time being, on his moving Earth.

The heavens challenge the astronomer, he noted in the opening paragraph of his *Brief Sketch*, to describe all the disparate motions of the divine bodies. His earliest predecessors in cosmology—here he saluted Calippus and Eudoxus from the fourth century B.C.—had embedded the Sun, Moon, and planets in a series of concentric spheres surrounding the Earth. The first astronomers envisioned these spheres as solid, invisible structures, each carrying a single planet. But these simple concentric spheres could not account for the planets' periodic brightening in the sky, as though they were drawing nigh. Later sages favored eccentric spheres, with centers *near* but not *at* the Earth, to make up the difference in brightness, and they tipped each sphere's axis at a slightly different angle, to allow the planets to bob up and down within the band of the zodiac. No single sphere, however, could embrace a planet's periodic reversals in direction. Anyone who watched the wanderers night after night saw them now and again slow down, stop, then backtrack with respect to the background stars for weeks or months on end, only to stop and start up again the way they had gone before—fluctuating in brightness all the while. To accommodate this loop-the-loop behavior, some astronomers imagined the heavenly spheres as clearly defined lanes, within each of which a single planet reigned. Inside the sphere of Mars, for example, the planet rode around on one or more subsidiary spheres, called epicycles, whose combined motions accounted for its constantly changing position in the sky. The undisputed master of this balancing act, Claudius Ptolemy, flourished in Alexandria around A.D. 150.

Ptolemy coped so effectively with heavenly complexity in the second century that he remained the reigning authority in the sixteenth. By following Ptolemy's instructions and using his tables, an astronomer could approximate the position of any planet at any time, past or future. As though to commemorate Ptolemy's magnificent achievement, his book came to be called by the first word of its title in Arabic translation, *Almagest*—"The Greatest"—instead of the more modest Greek name the author gave it, *Mathematike syntaxis*, or "Mathematical Treatise."

Copernicus revered Ptolemy as "that most outstanding of astronomers." At the same time, he objected to the way Ptolemy violated the basic axiom of astronomy, which held that all planetary motions must be circular and uniform, or composed of circular and uniform parts. Ptolemy conformed his Earth-centered ideology with his accumulated data on planetary speeds and positions by allotting each heavenly sphere a so-called equant—in effect a second axis of rotation, off-center from the true axis. Although astronomers deemed it impossible for any sphere to rotate uniformly about an off-center axis, they overlooked the infelicity because Ptolemy's technique worked on paper to give good predictive results. Copernicus's mind, however, "shuddered" at the thought.

Clinging to a purer ideal, as he explained in the *Brief Sketch*, Copernicus had sought a new route to Ptolemy's results, without committing Ptolemy's crime of violating the principle of perfect circular motion. Along the way to his new solution, for reasons Copernicus chose not to elaborate, he had nudged the Earth from its accustomed resting place at the hub of the universe, to put the Sun there in its stead. He could well have restored the heavens to uniform circular motion without this drastic reordering of the heavenly bodies, but once the new configuration occurred to him, the configuration itself became paramount.

"All spheres surround the Sun as though it were in the middle of all of them, and therefore the center of the universe is near the Sun," he wrote. "What appear to us as motions of the Sun arise not from its motion but from the motion of the Earth and our sphere, with which we revolve about the Sun like any other planet."

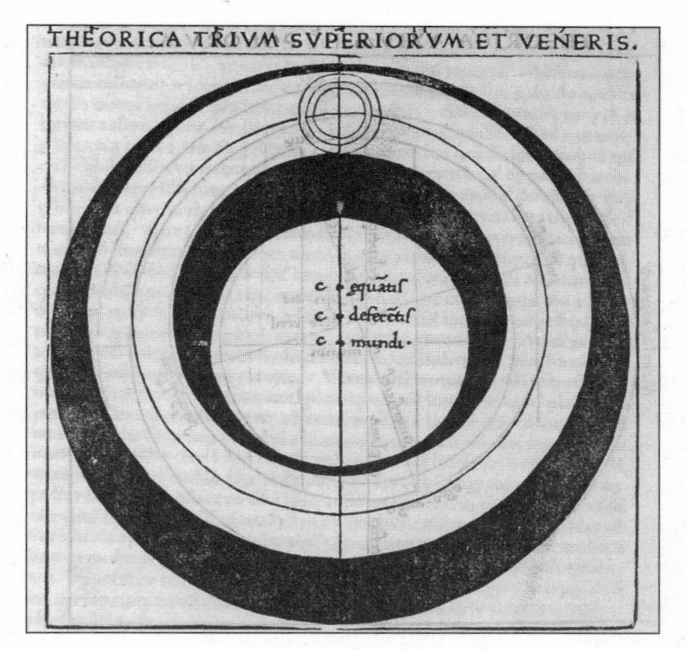

THE HEAVENLY SPHERES

Each planet traveled in its own sphere, or region of the heavens. As depicted in this image from Theoricae novae planetarum, *published in Nuremberg in the year of Copernicus's birth, the sphere has a three-dimensional quality, high and wide enough to embrace a planet's wanderings. Within the sphere, the principal path of the planet's motion traces a thin circle centered on a point labeled* c. deferentis. *It is an eccentric circle, since the Earth (the center of the universe) is located at the point just below,* c. mundi. *The top label,* c. aequantis, *indicates the equant point, from which the planet's motion appears uniform.*

With a wave of his hand, he had made the Earth a planet and set it spinning. In fact he saw it propelled in three ways, on three circular routes. The first spirited the planet around the Sun every year. The second twirled it daily, producing the heavenly fireworks of sunrise, sunset, and what he called the "headlong whirl" of stars through the night. The third motion slowly wobbled the poles over the course of the year, to account for the direction of tilt for the Earth's axis.

"Whatever motion appears in the sphere of the fixed stars belongs not to it but to the Earth," he continued. "Thus the entire Earth, along with the nearby elements"—meaning the oceans and the air—"rotates with a daily motion on its fixed poles, while the sphere of the fixed stars remains immovable and the outermost heaven."

Even Ptolemy had once conceded it might be easier in theory to let the Earth spin than to expect the whole firmament to wheel fully about every twenty-four hours—except that the idea of the Earth's turning was "utterly ridiculous even to think of."

As soon as Copernicus called for the Sun and the Earth to swap places, the planets snapped into a logical new order. They arrayed themselves outward from the Sun according to their speed of revolution, so that Mercury, long observed to be the fastest, was also the closest, followed by Venus, then Earth, Mars, Jupiter, and finally Saturn, the slowest. In the Earth-centric view, neither observation nor theory had ever settled the question as to which planet lay just past the Moon—whether Venus or Mercury—or whether the Sun orbited before, between, or beyond those two. Now he knew. Everything fit. No wonder the beauty of the system prevailed over the absurdity of the Earth's motion. He hoped his own conviction would convince others to see the spheres his way, but offered no proofs at this point. He had decided, he said, "for the sake of brevity to leave the mathematical demonstrations out of this treatise, as they are intended for a larger book." Then he proceeded to count and clarify all the individual planetary motions, arriving, in the final paragraph of the *Commentariolus*, at the grand total: "Mercury runs on seven circles in all; Venus on five; the earth on three, and round it the moon on four; finally Mars, Jupiter, and Saturn on five each. Altogether, therefore, thirty-four circles suffice to explain the entire structure of the universe and the entire ballet of the planets."

Copernicus surely anticipated ridicule from his contemporaries. If the Earth rotated and revolved at great speed, they could argue, then anything not nailed down would go flying. Clouds and birds would be left behind. Moreover, his fellow astronomers could insist the Earth truly belonged at the center—not because humanity's home deserved any special importance in the cosmic scheme, but because heavy, earthy things fell to rest there, and because change and death befell Earth's inhabitants. The Earth represented the pit, not the pinnacle, of Creation. Therefore one dare not shove the Sun—"the lamp of Heaven," as many called it—into the Hell hole at the center of the universe.

Several Islamic astronomers of the thirteenth and fourteenth centuries had found fault with Ptolemy for the same reasons Copernicus did. Nasir al-Din al-Tusi and Ibn al-Shatir, for example, managed to adjust Ptolemy's circle violations without requiring the Earth to turn or abandon its central place. Copernicus used some of the same mathematical devices in his revision of Ptolemy, but reached his own singular conclusions about the Sun's centrality, the Earth's mobility, and the grandiose inflation of the cosmos required by his design.

If the Earth trekked all the way around the Sun, as he maintained, then two neighboring stars should appear now slightly closer together, now farther apart over the course of the year. Yet the stars never displayed any such displacement, or "parallax." Copernicus got around the absence of parallax by supposing the stars too far away to reveal it. He increased their distance more than a hundredfold—so remote that the Earth-Sun separation shrank by comparison to the point of insignificance. "Compared to the great height of the sphere of the fixed stars," he averred, "the distance between the Sun and the Earth is imperceptible." The enormous chasm that suddenly yawned open between Saturn and the stars did not trouble Copernicus, as he had a ready explanation for it in the Creator's omnipotence: "So vast, without any question, is the divine handiwork of the most excellent Almighty." Beyond the periphery of the stars, God and His Angels hovered in the invisible heavens of the Empyrean.

After completing the *Commentariolus* around 1510, Copernicus began the slow work of elaborating his theory. The thirty-four circles of the planetary ballet now required exact design specifications, such as the radius of each one, its rate of rotation, and its spatial relationship to the other thirty-three. He could calculate many of these hundred-plus parameters using time-honored methods and tables. Then he would test the values by making his own observations.

The chapter, however, had other expectations of him.

In November 1510 Copernicus and fellow canon Fabian Luzjanski, who had studied with him in Bologna, went on an important mission to the chapter's southern provinces. There they accepted the large sum of 238 marks—a full year's revenue from the labor of peasants on Church land—for safe transport back to Frauenburg. Given that the Teutonic Knights were regularly and ruthlessly robbing the population of Varmia, the two couriers traversed the wooded, hundred-mile route home in constant danger of being waylaid and relieved of their cargo of coins. (Paper money had not yet come into circulation in Europe.) When they reached Frauenburg without incident, they distributed the funds among the canons according to custom.

The next November, in 1511, the chapter named Copernicus its chancellor, charged with overseeing the financial accounts and composing all official correspondence. The pace and volume of that correspondence quickened at the sudden death of his uncle the bishop on March 29, 1512. A week after Lukasz Watzenrode's passing, the canons met on April 5 to elect his successor. They voted unanimously for their own Fabian Luzjanski—all except Luzjanski himself, who wrote another's name on the ballot. The next day the canons gathered again and chose Tiedemann Giese to conduct the

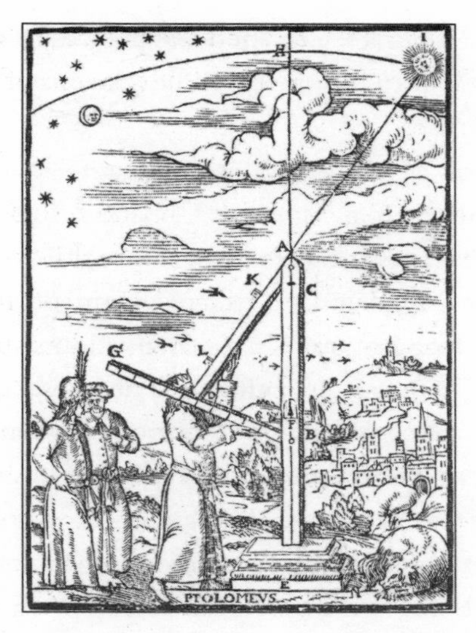

ASTRONOMICAL INSTRUMENTS

Between Frauenburg and Rome—the northern and southern limits of his lifetime travel—Copernicus could see most of the same thousand stars that earlier astronomers of Egypt, Babylon, Greece, and Persia had observed. He measured each star's celestial latitude and longitude to create a stellar catalog, which he published in On the Revolutions, *Book II, chapter 14. He also tracked the positions of the planets against the background of the stars. With his wooden triquetrum, like the one pictured here, he could gauge a body's altitude by sliding the hinged bar until its peepholes framed the planet or star, and then reading its elevation from the calibrated lower scale.*

necessary confirmation negotiations with the Vatican. By June 1 the chapter needed two more spokesmen to contend against the king's objections to their designated bishop. King Sigismund found no particular fault with Luzjanski; he simply preferred to install his own candidates in such positions. The Rome-Krakow-Varmia wrangle over the bishopric wore on through the summer and fall. On December 7, under a new agreement, Sigismund at last accepted Luzjanski, in exchange for the right of final approval over all future bishop selections. In addition, he insisted the entire chapter must pledge an oath of allegiance to the Crown, which they did on December 28, confident the king would honor, in return, his promise of royal protection.

Only one canon failed to sign the new agreement and swear his loyalty to the Polish king. This was Copernicus's brother, Andreas. The chapter had released him from all responsibilities in Frauenburg when he developed leprosy and, fearful of contagion, forced him to quit the region before Bishop Fabian's formal investiture. They could not strip him of his canonry, which carried a lifetime tenure, but death would do that soon enough. Not even Doctor Nicolaus could cure the biblical curse of this painful and disfiguring disease. Already, eager contenders vied to become Andreas's "coadjutor"—the person legally empowered to discharge his duties so long as he lived, and assume all his entitlements later, after his death. To a man, every canon could name some deserving relative qualified to fill this post. Naturally King Sigismund also had nominees in mind.

As Andreas left for Italy, in search of whatever solace he might find, Copernicus accepted a new line of duty as overseer of the chapter's mill, bakery, and brewery. These establishments provided bread and beer to the canons—and also served the peasants, for the price of dues, which Copernicus would need to collect from them.

On March 31, 1513, according to the Varmia ledger of accounts, "Doctor Nicolaus has paid into the treasury of the Chapter for 800 bricks and a barrel of chlorinated lime from the Cathedral workyard." With these materials he built a level platform in a garden near his curia. By now he had traded the first residence he had been given for this new one, which must have been better situated for his purposes. The large paved patio, or "pavimentum," as he called his construction, provided an unobstructed view of the sky and a solid footing for his astronomical instruments. He owned three with which he took the heavens' measure: a triquetrum, a quadrant, and an armillary sphere. None of these devices contained lenses or sharpened his vision in any way. Rather, they functioned as surveyor's tools, to help him map the stars and trace the paths of the Moon and planets.

In the spring of 1514, taking advantage of a redistribution of the chapter's property, Copernicus purchased living quarters inside the cathedral complex. While holding fast to his estate and pavimentum on the outside, he paid 175 marks for a spacious, three-story tower at the northwest corner of the fortification wall, complete with kitchen and servant's room. The top floor let in light through nine windows and gave out on a gallery, but he still preferred to observe from his patio platform. He stole hours from his sleep to stand watch out there, perched above 54° north latitude, on a forested hillside where the air hung heavy with mists off the Vistula Bay.

"The ancients had the advantage of a clearer sky," he wrote in his own defense. "The Nile, so they say, does not exhale such misty vapors as those we get from the Vistula." At the site of Ptolemy's fabled observatory on the Nile, with its near-tropical climate, the planets climbed almost straight up from the horizon, instead of loitering along the tree line, and rode high in the sky, easily sighted through countless cloudless nights.

Everything Copernicus knew about Ptolemy when he prepared the *Commentariolus*, he learned from an abridged interpretation of Ptolemy's work, called the *Epitome of Ptolemy's Almagest*, published in Venice in 1496. Now, as he launched his own serious research project to revise astronomy, the full text of Ptolemy's *Almagest* became available for the first time in a printed Latin translation. Copernicus consumed his copy, covering its margins with notes and diagrams.*

Copernicus found in the *Almagest* a model for the book he wanted to write, in which he would rebuild astronomy in a framework as

* Medieval scholar Gerard of Cremona (1114–1187) compiled this edition from several Arabic translations of the original (lost) Greek text. Gerard is said to have completed the work at Toledo in 1175, but its publication waited half a century after the invention of movable type, to be issued in Venice in 1515.

impressive and enduring as Ptolemy's. Meanwhile the *Commentariolus*, his prequel to *On the Revolutions*, was already making his name as an astronomer. This growing recognition no doubt accounted for the invitation Copernicus received from Rome to consult on calendar reform. The Julian calendar then in use, established by Julius Caesar in 45 B.C., had overestimated the length of the year by several minutes. Over time, the small founding error amounted to a gain of almost a day per century. Easter was on its way to becoming a summer holiday, and the Church's other movable feasts were falling similarly out of step with the seasons. Therefore Pope Leo X, as part of his agenda for the Fifth Lateran Council, called on theologians and astronomers of all nations to help correct the flaw. Copernicus duly submitted his comments to Paul of Middleburg, the Bishop of Fossombrone, who coordinated the calendar effort for the duration of the council, from 1512 to 1517. The bishop listed the contribution from the Polish canon in his official report. Unfortunately he did not describe even the gist of Copernicus's suggestion, and later, even more unfortunately, Copernicus's original letter was lost.

CHAPTER 3
Leases of Abandoned Farmsteads

Stenzel the herdsman took possession of 3 parcels, from which Hans Calau ran away. Stenzel got one ox, 1 cow, 1 piglet, 2 sacks of rye seed, nothing else. And I promised to add 1 horse.

—FROM AN ENTRY BY COPERNICUS IN THE VARMIA LEDGER, APRIL 23, 1517

LONG WITH HIS NAME and his faith, Copernicus inherited his country's long-standing conflict with the Knights of the Teutonic Order. His father had fought them hand to hand in Danzig and Torun, while his maternal grandfather, a Torun alderman, floated loans to finance the city's sporadic warfare with the order. As a boy, Copernicus loitered among the ruins of the knights' citadel in the city of his birth.

The knights first arrived in Torun early in the thirteenth century, fresh from bloodying Jerusalem during the Crusades. Several Polish dukes and princes invited them in, to control unruly elements throughout the province known as Old Prussia. With free rein and a heavy hand, the knights subdued the Balto-Slavic tribes that troubled the landed gentry, and converted the pagans to Christianity. They crusaded for five decades through territory they came to regard as their own—despite the prior claims of their noble hosts.

The knights' brutish ways chafed against the interests of the rising merchant class and town burghers. Around the year 1280, when Torun joined the German commercial cooperative called the Hanseatic League, the knights established grand new headquarters to the north, at Marienburg on the Nogat River. This sprawling castle and other Teutonic forts along the waterways—together with the port of Danzig, which they seized in 1308—made the knights the gatekeepers of the Baltic Sea. For the next hundred years they tempered their marauding with domination of the amber trade. The "Great War" they declared on Poland in 1409 went badly for them, however, because the scattered princes united against them under a strong new king. After this defeat, the knights' strength gradually diminished.

In 1454, around the time the elder Niklas Koppernigk moved to Torun, the residents of the city rose up against the order. The first clash of this "Thirteen Years' War" destroyed the knights' founding fortress. The final coup, delivered with the 1466 Treaty of Torun, deprived them of the western half of their realm in Old Prussia. Torun henceforth belonged to "Royal Prussia," officially annexed to the Kingdom of Poland. King Kazimierz IV occupied the knights' Marienburg castle for a time, but soon removed to the traditional royal seat at Wawel Castle in his native Krakow.

The knights retreated to the east, where they continued to rage against their Polish neighbors. Varmia particularly galled them. Even its geography provoked insult—the way this little bubble of Royal Prussia intruded into eastern Prussia by a narrow neck, then swelled within those borders. Bishop Watzenrode had warded off the order's aggression throughout the twenty years of his heyday. But Bishop Luzjanski lacked Watzenrode's power to command, and proved a poor match for young Albrecht von Hohenzollern, thirty-seventh Grand Master of the Knights of the Teutonic Order.

Albrecht was only twenty years old when the knights chose him as their leader in 1511. He had been bred for a Church career and already held a canonry at the Cathedral of Cologne. In addition to his Catholic devotion, his parentage appealed mightily to the order's liking: Albrecht's father—the German prince Friedrich I, Margrave of Brandenburg-Ansbach—ruled a choice piece of the Holy Roman

Empire; his mother, Princess Sofia of Poland, was King Sigismund's sister. Albrecht embodied the knights' every best hope of regaining their former luster, their former territory, and sovereignty over Prussia. True to that vision, Albrecht grew into his role with a vengeance, courting allies in Germany and Moscow as he braced for more war with Poland.

In mid-July of 1516, in Elbing, a town near Frauenburg, the Teutonic Knights robbed a citizen and maimed his hands. The Varmia Chapter sent a posse to give chase across the border into the knights' part of Prussia, where their guards caught one of the culprits and took him into custody. But Grand Master Albrecht demanded his subject's return. Then he retaliated with further attacks across Varmia. On July 22, Tiedemann Giese, who had succeeded Copernicus as chancellor, wrote up the canons' concerns in a desperate appeal to King Sigismund, beseeching him for the protection he had promised.

The uneasy standoff with the knights still prevailed the following November, when the Varmia Chapter elected Canon Copernicus to

Albrecht of Prussia, Grand Master of the Knights of the Teutonic Order.

administer its vast landholdings in the south. This post, which rotated among the sixteen canons, separated the current officer from the rest of the chapter by many miles and saddled him with new responsibilities.

Time and custom had divided the diocese of Varmia and its ninety thousand inhabitants into nine districts. The bishop personally owned six of these, including Heilsberg, where his palace stood. The other three belonged communally to the chapter: Frauenburg on the northern coast, the official seat and site of the cathedral; Melsack, its contiguous neighbor; and Allenstein at the southern extreme. Between them, Melsack and Allenstein contained 150,000 acres of fertile fields and pastures that fed the Varmia canons and generated their comfortable annual incomes. Keeping the land productive meant keeping it tenanted by peasants who shouldered all the hard labor of farm work—a personnel problem that would preoccupy Copernicus throughout his three-year term of office as administrator.

Immediately upon his election, he left his curia in Frauenburg for the chapter's southern headquarters. He had lived in one lost fortress of the Teutonic Knights when he served his uncle at Heilsberg, and now he moved into another: Allenstein Castle on the meanders of the Lyna River. He assumed his new post on St. Martin's Day, November 11, 1516, which, according to the Church calendar, marked the first day of the new ecclesiastical year 1517.

The picture of Copernicus that normally comes to mind—the solitary figure, cloistered with his books in a monkish study carrel, or ascending some parapet to implore the night sky—all but falls apart at Allenstein. His time there plunged him among the people as a sympathetic party to all their mundane concerns.

The peasants under his charge lived in huts, in poverty, and in dread of the knights who raided their villages. They paid the chapter a fee of one Prussian mark per year, per parcel of land, for the privilege of plowing, sowing, and harvesting—though the Church also claimed the bulk of their harvest as revenue. In a sense a peasant owned his land, because he could trade it to someone else, or pass it down to his children. But in fact the chapter lorded over everything, and kept track of all exchanges of parcels by recording their locations in official ledgers. On a clean page in one of these books, the new administrator wrote, "Leasing of Farmsteads by me, Nicholas Coppernic, A.D. 1517."

Duty called him first to Jonikendorf, where he approved Merten Caseler's takeover of three parcels of vacant land. The former tenant, Joachim, had been hanged for thievery. On account of his crimes, or his punishment, Joachim had failed to sow his fields, and so Copernicus waived Merten Caseler's rent for the whole year. He also noted the several assets that accompanied the three parcels: "He got 1 cow, 1 heifer, an ax and a sickle and, as for grains, a sack of oats and barley for the sowing omitted by his predecessor." Copernicus datelined his description of this business "weekday 4"—meaning Wednesday—"10 December 1516." After that he wrote, "In addition, I promised him 2 horses."

Copernicus traveled on horseback among the region's 120 villages, often accompanied by his servant, Wojciech Szebulski, or his errand boy, Hieronim, both of whom he named in the ledger as frequent witnesses. He alone, however, decided every case, and his word spoke for the whole chapter.

"Bartolt Faber of Schonewalt took possession of 1½ parcels, sold by Peter Preus, who is very old. As regards these parcels, Bartolt will give the overlord ½ mark as rent for the half-parcel. But as regards the other parcel, the Chapter graciously donated 1 mark to the aforesaid Peter for life." In other words, Copernicus allowed Bartolt Faber to rob the chapter (the overlord) to pay the aged Peter Preus an annuity through his declining years. "After his death the entire rent will revert to the overlord. Done on the second weekday after Laetare [March 23], 1517, in the presence of Wojciech, my servant, and Hieronim, etc."

Similarly, when Alde Urban, "who is aged in fact and in name"—and who had "no sons"—felt compelled to cede one of his parcels, Copernicus granted him exemption from payments on the rest of his holdings. Jan of Vindica, on the other hand, received no such exemption when he took possession of four parcels. Apparently Copernicus judged Jan well provided for by an uncle on his mother's side, Czepan Copetz, who had worked that land to his dying day and left upon it "4 horses, 1 colt, 4 cows, 6 pigs, 1 leg of pork, 1 sack of rye, 1 sack of flour, ½ sack of peas, 4 sacks of barley, 5 sacks of oats, 1 large kettle, 1 wagon, iron plowshares, 1 ax, 1 scythe."

In Voytsdorf, Copernicus encountered another family with a good uncle likely to stir his memories of the old bishop: "Gregor Knobel adds to his 2 parcels 1 more parcel that belonged to Peter Glande, who died in a fire. Gregor is the guardian of his brother Peter's sons, who are minors, and promises to satisfy them when they are grown up."

Copernicus's own grown brother now wandered alone somewhere in Italy, a leper, shunned by everyone, as the nerves in his skin slowly disintegrated. The last communication from Andreas, by proxy the previous February, acknowledged that he had received his share of Uncle Lukasz's estate. More than likely, those funds would see him through to the end.

"Hans Clauke has 2 parcels for which he was bound by hereditary payments to the church in Berting. As a man incapacitated for a long time, he sold those parcels to Simon Stoke with my permission. Done on 4 May."

If Copernicus lent his medical skills to any of the sick or elderly peasants, he did not note such treatments in the ledger. Death—whether by hanging, by fire, by illness or old age—caused the usual degree of attrition in the population. Desertion also took its toll.

"Jacob Wayner, who with his wife ran away last year, has now been brought back by the overseer," Copernicus wrote on August

Copernicus holds a lily of the valley, an early Renaissance symbol of a medical doctor (probably because of the flower's association with the god Mercury, whose snake-entwined caduceus promoted healing), in this wood-block portrait by Tobias Stimmer.

2, 1517. The peasants' hard lot caused many of them to flee in search of a better life. More than one quarter of the cases Copernicus recorded made reference to land made vacant because Simon—or Martzyn or Cosman—had run away. The village overseer typically pursued these fugitives on behalf of the chapter and returned them to work, lest the land lie fallow or, worse, revert to its wooded state, in which case new cultivators would have to be bribed to clear and resow.

"Jacob took possession of one parcel," Copernicus's account continues, "from which death removed Caspar Casche. The building is in ruins, and the parcel is of little value, and for that reason was abandoned by Caspar's heirs and guardians. When Jacob took possession, I gave him one horse, a quarter of the previously planted millet, and exemption from the next annual payment." Copernicus also named Michael Wayner, brother of the runaway, as "his guarantor in perpetuity"—to guarantee that Jacob would never again run away.

"Gregor Noske took possession of 1½ parcels, from which Matz Leze ran away because he was suspected of thievery."

Land changed hands among the peasants in every month of the year—from "the antepenultimate of January" and "the Sabbath before Palm" to "the day of Peter and Paul," "the feast of Michael," "St. Cecilia's," and "the day of the 11,000 Virgins."

As he alternated between saintly and standard designations for dates in his record keeping, Copernicus kept up his lone struggle to define the true duration of the year. His invited remarks to the Lateran Council about the problematic Julian calendar had probably lamented astronomers' ignorance of the year's exact length. With or without calendar reform, Copernicus still needed to ascertain this fundamental parameter. The length of the year defined the Earth's orbit around the Sun—or, as other astronomers believed, the Sun's orbit around the Earth—and pertained to almost every calculation in the heliocentric or any other planetary theory.

"Petrus, a herdsman in Thomasdorf, took possession of 2 parcels, which are vacant because Hans ran away."

Copernicus fashioned a new yardstick for the year in an open loge on the south face of Allenstein Castle, just outside his private apartment. Laying white stucco over the ruddy bricks, he painted the grid of a sundial onto the smoothed surface. The lines and numbers must have been blue and red when new, though only a hint of color survives in the faded dial fragment still clinging to the castle wall. Underneath, either on a table or the floor, he set a mirror—or maybe he used a bowl of red wine—to catch the Sun's reflection and throw it up to the dial, where he charted the changing solar altitude through the seasons.

"Jacob of Jomendorf took possession of 2 parcels, which were sold to him with my permission by Marcus Kycol, who is very old."

The Sun reaches its highest point at the summer solstice, which occurs on the longest day of the year, and a year may be gauged by the lapse of time between one summer solstice and the next. Or one could measure the time passing between one year's vernal equinox—when the Sun crosses the Equator at the start of spring, dividing the day into equal halves of dark and light—and the vernal equinox that follows. The equinox proved easier than the solstice for Copernicus to capture, because the Sun's position changes more dramatically day to day during the run-up to equal day and night than it does near the longest day of the year. Still, determining the exact moment of the equinox challenges even the most diligent observer. In some years, the moment defies observation altogether if it occurs during the night or twilight hours of the given day.

Copernicus worked around the natural obstacles by making a series of noon observations over a period of several days before and after the anticipated event, and then interpolating the time. His calculation gave him the year's duration down to minutes and seconds in an age when no clock could shave moments so closely. He repeated the process annually, pooling his figures to improve their accuracy. He also factored in a few results from Ptolemy, to further increase his baseline, and he adopted Ptolemy's technique of reckoning dates by the reigns of ancient rulers. Thus Copernicus recalled observing an autumnal equinox at Frauenburg "in the year of Our Lord 1515 on the 18th day before the Kalends of October, but according to the Egyptian calendar it was the 1840th year after the death of Alexander on the 6th day of the month of Phaophi, half an hour after sunrise." Despite its awkward phraseology, the Egyptian calendar appealed to Copernicus's contemporaries because of its consistency: The list of kings stretched all the way back to the eighth century B.C., and every year consisted of twelve identical months, thirty days each, plus an extra five days tacked on at the end—with no leap years. A sixteenth-century date converted to Egyptian style allowed easy computation of the time elapsed since any similar observation by Ptolemy.

"Jacob has 2 parcels and sold them with my permission to Lorenz, the overseer's brother."

The coins the peasants put down on their transactions were a mishmash of old and new currencies, both Prussian and Polish. The Teutonic Knights had been minting Prussian marks in the region since the thirteenth century, but at the start of the Thirteen Years' War in 1454, King Kazimierz extended minting privileges to the cities of Torun, Elbing, and Danzig. The burghers then turned out their own Prussian coins at an enthusiastic rate, in the familiar denominations: marks, skoters, groats, and pence. In the absence of anything resembling national standards or official exchange rates, however, the intrinsic value of a mark—the volume of silver it contained—varied from mint to mint. Even the same mint might shift the balance of silver to copper in its specie on a whim. Thanks to a suspiciously diminishing proportion of silver in successive issues, a new mark weighed less than an old one, while pretending to equal the old value. Copernicus proved the weight difference by comparing coins in a balance pan. He knew that canny citizens were taking advantage of the discrepancy by spending the new coins and hoarding the old ones, which they would take to the goldsmith to be melted down for the greater worth of their metal content. Other abuses, such as nipping off bits of coins' edges, also contributed to the currency's debasement. Sometimes the pennies the peasants paid toward their rent, despite a full quotient of the

Martin Luther, "the great Reformer," as depicted by Lucas Cranach the Elder.

proper alloy, had been pinched and handled through such long use as to be worn thin.

Aware of all these ills, Copernicus spent part of his first summer in Allenstein penning a private reflection on his fears for the state of the currency. He completed the *Meditata*—his Latin meditation on the money problem—on August 15, 1517, and then circulated it among a few chosen associates, much as he had done with the *Brief Sketch*.

At the same time as Copernicus listed these financial concerns, the priest and theology professor Martin Luther in Wittenberg also drew up a list. Luther's list enumerated his many complaints against the Catholic Church, contesting its sale of indulgences as tickets to salvation. "When the money clinks in the box," Luther had heard some mercenary clerics claim, "the soul springs up to Heaven." Like Copernicus, Luther directed his so-called 95 Theses to a small, select group of acquaintances. But while Copernicus's money advice drew a polite response that hardly distracted him from his duties, Luther's outrage lit a fire that soon filled the public squares.

"Voytek, who has 2 parcels in the same place, took possession of two additional parcels, which have been abandoned for a long time on account of the flight long ago of Stenzel Rase. Voytek will pay the next annual rental."

"Lurenz, having bought the tavern in Branswalt, with my consent sold 4 parcels."

In November of 1518, word reached Copernicus that the sickly Andreas had succumbed at last to the final stages of leprosy and left this world. As he mourned the death of his brother, his friend Tiedemann Giese lost his two sisters to an outbreak of plague in Poland.

"Stenzel Zupky took possession of 2 parcels, which Matz Slander with my permission sold to him for 33 marks."

Some of the local civic authorities who read Copernicus's currency essay deemed it worth discussing at a regional assembly. Copernicus agreed to translate the text, for the benefit of representatives from Danzig, into German (still the official language in that city, despite its fealty to the Polish king). He finished the revised version by the end of 1519, when his term as administrator ended, and looked forward to seeing his suggestions implemented for the standardization of coins and the improvement of minting practices. Within weeks of his return to Frauenburg, however, the long-threatened war with the Teutonic Order broke out. On December 31, Albrecht invaded Braunsberg, the largest town in Varmia. Copernicus rode the six miles from Frauenburg to try to reason with Albrecht, but after two days of effort as the bishop's emissary, January 4 and 5, all he won from the grand master was a promise of safe conduct through the region should he wish to resume negotiations in the future. He went home in defeat.

A fortnight later, on January 23, 1520, Albrecht's knights attacked Frauenburg. They sacked and burned it. Only the walled cathedral complex, protected by a phalanx of Polish soldiers, escaped destruction. Copernicus's curia outside the walls was reduced to rubble and ash. His pavimentum, too, lay in ruins.

Chapter 4
On the Method of Minting Money

Coinage is imprinted gold or silver, by which the prices of things bought and sold are reckoned according to the regulations of any State or its ruler. It is therefore a measure of values. A measure, however, must always preserve a fixed and constant standard. Otherwise, public order is necessarily disturbed, with buyers and sellers being cheated in many ways, just as if the yard, bushel, or pound did not maintain an invariable magnitude.

—From Copernicus's revised money essay, 1522

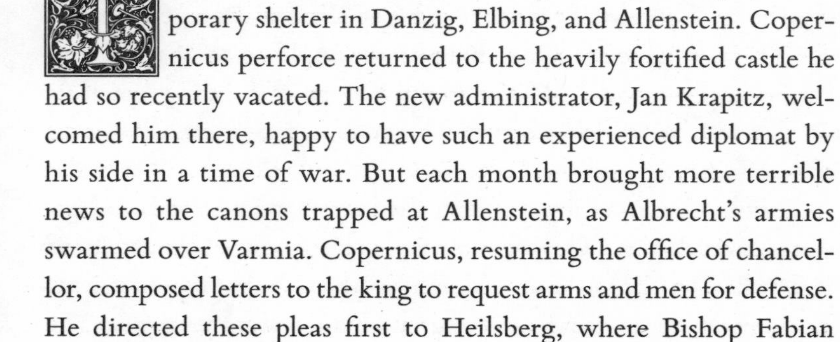

THE CANONS FLED the torched city of Frauenburg for temporary shelter in Danzig, Elbing, and Allenstein. Copernicus perforce returned to the heavily fortified castle he had so recently vacated. The new administrator, Jan Krapitz, welcomed him there, happy to have such an experienced diplomat by his side in a time of war. But each month brought more terrible news to the canons trapped at Allenstein, as Albrecht's armies swarmed over Varmia. Copernicus, resuming the office of chancellor, composed letters to the king to request arms and men for defense. He directed these pleas first to Heilsberg, where Bishop Fabian signed them and sent them on to Sigismund in Krakow. Sometimes the enemy intercepted the desperate correspondence, and sometimes the king received it but could not comply. Even when he responded with reinforcements, the new recruits failed to rout the knights.

Confined to the castle, Copernicus continued the planetary

observations that clarified his picture of the universe. On February 19, 1520, his forty-seventh birthday, he judged Jupiter, at 6:00 A.M., to be 4°3' to the west of "the first, brighter star in the forehead of the Scorpion." Sometime in the spring, Jupiter would reach its annual opposition, to appear exactly opposite the Sun in Earth's skies. No one could see both bodies at once when that happened, but an experienced practitioner might fix the time by combining predictions from theory with observations over a period of months. Having thus begun his watch in February, Copernicus marked the moment of opposition at 11:00 A.M. on April 30. Jupiter was then moving in reverse, or "retrograde," as though backing away from the Scorpion's sting—and also making its closest approach to Earth. While other astronomers viewed the timing of these several events as coincidence, Copernicus linked them inextricably together as the consequence of planetary order: The Earth, being closer to the Sun, overtook the slower Jupiter once per year. In passing, it left the Sun to one side and Jupiter on the other, coming as close to Jupiter as ever it could. Jupiter itself never changed the direction of its movement at such times, but merely appeared to do so to observers on Earth as they sped by. The same logic governed the annual opposition of Saturn, which Copernicus tagged a few months later, at noon on July 13.

The movements of Jupiter and Saturn at this juncture raised alarm among astrologers. The two planets were heading toward their "Great Conjunction"—the close heavenly union that they consummated every two decades, always with momentous effects. The popular almanac by Johann Stoeffler and Jacob Pflaum foresaw in the upcoming Great Conjunction of 1524 "changes and transformations for the whole world, for all regions, kingdoms, provinces, states, ranks, beasts, marine animals, and everything that is born of the earth—changes such as we have hardly heard of for centuries before our time, either from historians, or from our elders. Raise your heads accordingly, Christian men."

On October 19, 1520, a detachment of knights surrounded Bishop Fabian's palace at Heilsberg and settled in siege there for weeks. Under these circumstances, the chapter turned Jan Krapitz out of office at the November election, though he had served only one year, and drafted Copernicus to replace him. On the day his second stint as administrator started, November 11, restless knights hovered within a day's ride of Allenstein, where only a hundred of the king's soldiers stood guard at the gates.

Copernicus filled some of his most anxious hours cataloguing the chapter's archives, which had been moved to Allenstein over the years for safer keeping inside the castle treasury. The whole embattled history of the diocese lived in these documents, going all the way back to the bull of Pope Innocent IV in 1243, defining the boundaries of Prussia, and the 1264 parchment on which Anselm, the first Bishop of Varmia, envisioned the raising of a great cathedral at Frauenburg. The several hundred items—bulls, treaties, grants, deeds, wills, testimonials, petitions—filled a chest of many drawers. Even as Copernicus sifted and re-sorted the legal materials with their ornate official seals, he wrote new letters of appeal, begging King Sigismund to bolster the forces protecting Allenstein's repository:

"Most Gracious Prince and Lord, Sigismund, by the Grace of God King of Poland, Grand Duke of Lithuania, Sovereign and Hereditary Lord of Ruthenia and Prussia, and our Most Gracious Lord," Copernicus addressed His Majesty on November 16. He described the awful details of the previous day's invasion of the nearby city of Gutstadt, now fallen to the knights, and expressed his willingness to die—as it seemed likely he would—in defense of Allenstein.

"For we are desirous to do what befits noble and honest persons, who are completely devoted to Your Majesty, even if we had to perish. All our possessions and ourselves we commend and entrust to Your Majesty's care." Meanwhile he continued methodically to list those possessions: "Document concerning the transfer of

In this 1520 painting by Stanisław Samostrzelnik, King Sigismund kneels beside the Bishop of Krakow to be blessed by St. Stanislaw, the patron saint of Poland.

the head of St. George from Heilsberg to Frauenburg Cathedral," "Document of the king of France concerning a gift of wood from the Holy Cross."

Sigismund's infantry marched in at the end of November. The Varmia canons took little comfort from the presence of the troops, however, and abandoned the castle in terror. Only Copernicus and

Canon Henryk Snellenberg stood their ground at Allenstein. They were there when the king's cavalry arrived in December, and even then the two canons did not relax their vigil. They faced their greatest test in January 1521, when Albrecht and an expanded army of his order demanded the castle's surrender. But then, in a sudden reversal, Albrecht merely ransacked the nearby villages and withdrew homeward toward Königsberg, having agreed to a provisional cease-fire. Still Copernicus would not stop shoring up the castle's defenses. He procured cartloads of the long guns called harquebuses from Elbing, and lead for shot, and food, and salt. Ready now for any sort of escalation, he received word of the April 5 treaty, signed in Torun, declaring a four-year truce.

Peace put Copernicus briefly back to work on the more mundane matters of administration. In May 1521 he oversaw the reassignment of land parcels made vacant "by the death of Michel the one-eyed," "by the beheading of Peter in Hoensteyn for plotting treason," and for a variety of other reasons. Both the peasants and the land had suffered war losses.

In June the chapter called Copernicus back to Frauenburg to restore normal order in the north, while his capable friend Tiedemann Giese took over at Allenstein. Giese's new term as administrator—this was his third—proved the most difficult of his career. Despite the peace treaty, the knights continued laying waste to Varmia. And yet, because of the peace treaty, the chapter could not fight back. Giese wrote petitions for redress of grievances. He made impassioned appeals for peaceful relations at summit meetings between the Prussian Estates and the Teutonic Order. Nothing, it seemed, would dislodge the knights from the city and environs of Braunsberg, which they had occupied since the start of the recent war. Still Giese persisted in his negotiations. Both the king and the bishop promised to support him at the summit planned for Graudenz in March 1522. Sigismund would send his emissaries, and Fabian would attend in person. In the event, however, the bishop was too ill to leave his bed, so he sent his physician, Doctor Nicolaus, in his stead.

Copernicus joined Giese in Graudenz, only slightly delayed by flooded bridges over the River Bauda that all but barred his way out of Frauenburg. On March 18, he stood with Giese before the assembled representatives and seconded his enumeration of the knights' abuses. Three days later, he delivered the treatise on coinage that he had conceived before the war, chastising the minting practices that had sent the currency into free fall.

"The worst mistake," he charged, "which is absolutely unbearable," is for the government to mint new coins—of inferior intrinsic worth, though pretending to equal value—while the old coins are allowed to remain in circulation. "The later coinage, always inferior in value to the earlier coinage, . . . constantly depressed the market value of the previous coinage, and drove it out."*

Copernicus compared the infusion of inferior coinage to the sowing of bad seed by a stingy farmer. The government, like the farmer, would reap exactly what it sowed, he said, since its practices damaged the currency as surely as blight ruined grain.

"Such grave evils, then, beset Prussian money and, because of it, the whole country," he continued. "Its calamities and decline benefit only the goldsmiths, who take the value of the money into their own hands."

To remedy the situation—"before a greater disaster!"—Copernicus recommended a consolidation of the mints, so that "only one place should be designated for the minting of money, not for a single city or under its emblem, but for the entire country." Mint no new money in the interim, he further counseled, and above all set strict limits

* Copernicus's realization that bad money drives good money out of circulation often goes by the name Gresham's Law, in honor of Sir Thomas Gresham (c. 1519–1579), a financial adviser to English royalty who made the same wise observation. The concept was also put forward by medieval philosopher Nicole Oresme and mentioned by the Ancient Greek playwright Aristophanes in his comedy *The Frogs*.

for the number of marks to be struck from a single pound of fine silver. Then, as soon as the new currency is introduced, prohibit the use of the old, so as to force the surrender of outdated coins for new ones—at a loss, yes, but only a slight one. "For, this loss will have to be suffered once, in order that it may be followed by many benefits and a lasting advantage, and that a single currency reform in 25 or more years may be enough."

His suggestions sounded even more relevant now that Sigismund wished to unify the disparate currencies of his kingdom. In order for the Polish Crown coins and Prussian monetary systems to be reconciled, a firm rate of exchange had to be established among them. Copernicus immediately produced an addendum to his treatise while still at the March 1522 meeting, outlining a specific plan to equalize the currencies. But the assembly ended without changing the status quo, either of coins in the realm or the knights ensconced in Varmia.

Grand Master Albrecht had condoned the minting of inferior Prussian coins from the moment he took over the Teutonic Order in 1511. Nevertheless, the high cost of war had put him on the brink of bankruptcy. Albrecht betook himself to Germany in 1522 to attend the Diet of Nuremberg, where he hoped to acquire new allies—and sufficient cause to break the peace with Poland. At the Diet, Albrecht met one of Martin Luther's disciples, a former Catholic priest named Andreas Osiander. As an avowed convert to the new evangelical Lutheran faith, Osiander set about convincing Albrecht to convert as well. Albrecht then went to Wittenberg to seek out Luther for further consultation. The now famous heretic, excommunicated by Leo X in 1521, likewise urged Albrecht to end his allegiance to the Catholic Church—and also his allegiance to the Order of Teutonic

Coins from the reign of Sigismund I.

Knights. Luther thought it preferable for Albrecht to appropriate the knights' side of Prussia as his own, find a wife to rule it with him, and raise a family dynasty to inherit his privileges. As Albrecht began to explore these intriguing avenues, his ten-year enemy, Fabian Luzjanski, the Bishop of Varmia, died of syphilis on January 30, 1523.

Copernicus, who had shunned the easy road to the bishopric that his uncle once paved for him, now found himself presiding at Heilsberg Palace. The chapter chose him to watch over all the lands of the diocese, including those appertaining to the bishop's see, until Fabian's successor was installed. Given the rancor that had surrounded the selection process after Bishop Watzenrode's passing, King Sigismund sent his envoys to Heilsberg in February to foil any preemptive election attempts by the chapter. Copernicus received these men with assurances that the canons would not only honor the king's right of nomination and approval, but also swear their allegiance to him anew under the next bishop, whosoever that might be.

On April 13, the chapter chose the king's favorite, Maurycy Ferber. Bishop-elect Ferber, a distant relative of Tiedemann Giese, belonged to a politically powerful family from Danzig, where one of his kinsmen currently served as mayor of the city. Pending papal approval of Ferber, Copernicus acted as de facto bishop all through the spring and summer of 1523. He struggled to restore law and order by ridding the region of recalcitrant knights and the rear guard of the Polish military. The very forces who had come to Varmia's defense now illegally occupied several villages and fortresses. They refused to leave until the king intervened. Following Sigismund's orders of July 10, all Polish commanders of troops squatting in the diocese finally relinquished their claims and decamped. The knights, however, remained.

In August the Moon turned red—not as a metaphor for blood or war, but actually and naturally, as the result of a total lunar eclipse.

Copernicus noted the first dip of the full Moon into the cone of the Earth's shadow at "2 and 4/5 hours past midnight," or 2:48 A.M. on August 26.

Traversing the shadow of the Earth, the Moon dimmed by degrees until fully immersed. Then, instead of disappearing in darkness, the eclipsed Moon daubed itself with the Sun's color: It glowed like an ember throughout the hour of totality, reflecting all the dusk and dawn light that spilled into Earth's shadow from the day before and the day ahead.

Copernicus never missed a lunar eclipse. No astronomer let such an opportunity slip, for the Moon in eclipse pinpointed celestial positions as no other phenomenon could. At such times the Earth's shadow became visible on the Moon's surface, and the center of that shadow indicated the location of the Sun—180° opposite in celestial longitude. With the Moon's current coordinates thus confirmed, one could also measure the distances of stars and planets from either the Sun or the Moon. "In this area," Copernicus remarked, "Nature's kindliness has been attentive to human desires, inasmuch as the Moon's place is determined more reliably through its eclipses than through the use of instruments, and without any suspicion of error."

Even with the help of "Nature's kindliness," the tilt of the Moon's orbit relative to the Earth's great circle limited the frequency of lunar eclipses to once or at most twice a year, though some years have none. After August 26, there would not be another total lunar eclipse till the end of December 1525.

At the moment of mid-eclipse, which Copernicus recorded on this occasion as 4:25 A.M., the Moon stood at opposition, yet stayed its course straight ahead. Unlike Jupiter or Saturn, the Moon never shifted into reverse at opposition—or ever, at any time—because the Moon, alone among all heavenly bodies, truly did orbit the Earth.

"In expounding on the Moon's motion," Copernicus wrote, with no apparent irony, "I do not disagree with the ancients' belief that it takes place around the Earth."

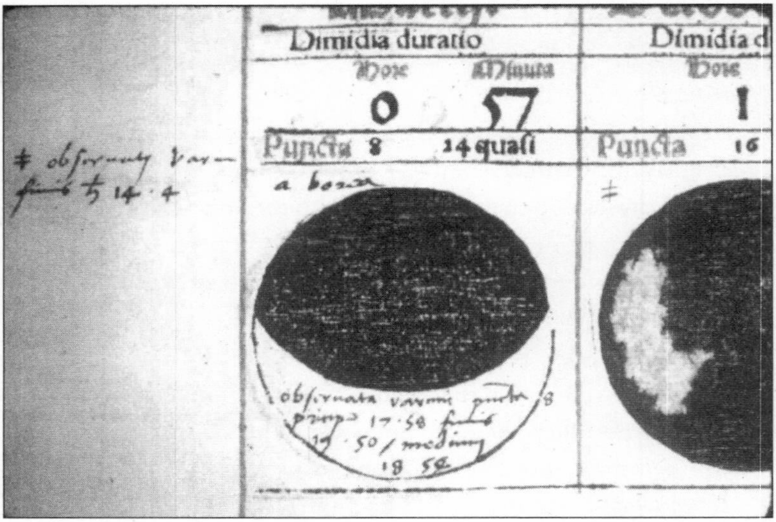

THE VALUE OF ECLIPSES

Johann Stoeffler's Calendarium Romanum magnum*, published in 1518, predicted eclipses for the years 1518 to 1573. Copernicus annotated his copy with his own observation notes between 1530 and 1541. The special alignment of Earth, Moon, and Sun at eclipse, called syzygy, provided a natural check on celestial positions. Copernicus witnessed both partial and total lunar eclipses, but only partial solar ones. Had he been able to travel to Spain or to the southern extreme of Italy for the April 18, 1539, event, he might have seen the Sun totally eclipsed.*

Ptolemy had reported in the *Almagest* how he derived the Moon's motion by tracking it through three eclipses of similar duration and geometry. Copernicus was following suit by observing his own three eclipses: one through the midnight hours of October 6–7, 1511; a second more recently, on September 5–6, 1522; and the third in the triad on this night, August 26, 1523. With these data, he meant to reroute the Moon.

On the path Ptolemy had charted centuries earlier, the Moon altered its distance from Earth so dramatically over the course of the month as to make it appear four times larger at its closest approach than at its most distant point. Observers never saw the Moon do anything of the kind, however. Its reliable diameter barely ever changed, yet Ptolemy and most of his followers ignored that glaring fact. Copernicus addressed the discrepancy by offering an alternate course that preserved the Moon's appearance.

On October 13, Bishop Ferber at last assumed his rightful position, freeing Copernicus to return to Frauenburg. The chapter elections in November named him chancellor once again, but he did not expect the duties of office to impede his astronomical researches or the writing of his book any longer. At fifty, he could only guess how much time remained for those pursuits, before the inevitable loss of his stamina, or his eyesight, or the clarity of his mind.

Chapter 5
The Letter Against Werner

Faultfinding is of little use and scant profit, for it is the mark of a shameless mind to prefer the role of the censorious critic to that of the creative poet.

—From Copernicus's *Letter Against Werner*, June 3, 1524

HE GREAT CONJUNCTION OF 1524 brought Jupiter and Saturn together in the sign of Pisces. Astrologers, who classified Pisces as a watery sign, predicted the dread disaster at conjunction would take the form of a mass drowning, indeed, a global inundation to rival Noah's flood. Every Jupiter-Saturn union blew an ill wind, but this one's evil potential drew added force from the number of other heavenly bodies convening with the main two. On February 19, Copernicus's birthday, the planets Jupiter, Saturn, Mars, Venus, and Mercury would all cluster together with the Sun in a grand sextuple conjunction, followed by a full Moon that night. Further proof of apocalypse derived from Pisces' rank order as the twelfth and final zodiac sign. Given that astrologers believed the world had begun under a multiplanetary conjunction in Aries, the first sign, surely it would end now under a repeat occurrence in Pisces, the last. The growth of both printing and literacy helped spread these dire prognostications so far and wide that people living in coastal regions took to the mountains. Some looked to their Bibles for instructions on how to build an ark.

February passed, and no floodwaters rose. Disbelievers scoffed at the astrologers, who held firm that waves—if not of water, then of religious dissent or political unrest—would yet wash over Europe. Had not the Great Conjunction of 1345 required two years to unleash the Black Plague?

Copernicus, who neither issued nor heeded astrological forecasts, chose this moment to pursue a bad debt. Canon Henryk Snellenberg, who had been his sole comrade in arms during the final defense of Allenstein Castle, went to Danzig and, as a favor, collected some money owed to Copernicus by his cousin on the city council there. But when Snellenberg returned to Varmia, he turned over to Copernicus only ninety of the hundred marks the astronomer's cousin had paid. Snellenberg repeatedly put off the reimbursement of the remaining ten marks, making one excuse after another over a period of months. When Copernicus finally confronted him, Snellenberg demanded written proof of the debt, and then dared his creditor to file suit for the sum still owed. Sufficiently grieved, Copernicus complained to Bishop Ferber.

"I therefore see that I cannot act otherwise," he wrote to his superior on February 29, "and that my reward for affection is to be hated, and to be mocked for my complacency. I am forced to follow his advice, the advice by which he plans to frustrate me or cheat me if he can. I have recourse to your Most Reverend Lordship, whom I ask and beseech to deign to order on my behalf the withholding of the income for his benefice until he satisfies me, or a kind provision in some other way for me to be able to obtain what is mine."

In comparison to the petulant but principled tone of his complaint against Snellenberg, the content of another letter Copernicus wrote that same year, on June 3, 1524, contained an invited analysis of such interest to the mathematics community that multiple copies of it circulated among his peers. Although terse and informal, the *Letter Against Werner* stands alongside the *Brief Sketch* and *On the Revolutions* as the third pillar of Copernicus's oeuvre in astronomy. He addressed it "To the Reverend Bernard Wapowski, Cantor and

Practica vber die grossen vnd manigfeltigen Coniunction der Planeten/die im jar M.D.XXiiij. erscheinen/vñ vngezweiffelt vil wunderparlicher ding geperen werden.

Auß Rö.Kay.May.Gnaden vnd Freihaiten/hüt sich menigklich/diese meine Practica in zwayen jaren nach zütrucken bey verlisung.4.Marck lötigs Golts.

GREAT CONJUNCTION OF 1524

The combined presence of Jupiter and Saturn—along with several other celestial bodies—in the twelfth zodiac sign, Pisces (the fishes), struck dread in the hearts of astrologers, who forecast the floodwaters shown pouring from the exaggerated sky-fish in this image from Leonhard Reynmann's Prognostication *for 1524.*

Canon of the Church of Krakow, and Secretary to His Majesty the King of Poland, from Nicolaus Copernicus."

Wapowski and Copernicus had attended the Collegium Maius in Krakow together as undergraduates in the 1490s. Possibly they developed their shared interest in planetary theory at that time, perhaps even in each other's company. Wapowski, who also went on to study law at Bologna, had later served several years in Rome with the Polish embassy, and now communicated with an international coterie of intellectuals. Copernicus alluded to the closeness of their long-standing friendship in his *Letter*'s first sentence.

"Some time ago, my dear Bernard, you sent me a little treatise on *The Motion of the Eighth Sphere* written by Johannes Werner of Nuremberg." Wapowski had sought Copernicus's opinion of this widely praised paper, which was published in 1522 along with several other recent essays by the same author. Copernicus hesitated before complying, however, because he found fault with Werner's thesis and was not at all sure he should say so. Now he excused himself to his old friend for the long delay.

"Had it been really possible for me to praise it with any degree of sincerity, I should have replied with a corresponding degree of pleasure." Unfortunately, the highest compliment he could offer—"I may commend the author's zeal and effort"—took him "some time" to muster. At first, he admitted, he feared arousing anger by expressing censure in writing. Better, perhaps, to say nothing at all against Werner than risk a negative backlash that might ruin any chance of a favorable reception for his own work.

"However, I know that it is one thing to snap at a man and attack him, but another thing to set him right and redirect him when he strays, just as it is one thing to praise, and another to flatter and play the fawner." In the best spirit of correcting a fellow astronomer's misstep, then, he would share his thoughts. He did not know that Werner, a clergyman at a Nuremberg infirmary, had died of the plague in 1522 while his papers were still on press. "Perhaps my criticism may even contribute not a little to the formation of a better understanding of this subject."

The "eighth sphere" of Werner's title spun the stars. They were all embedded in it, like jewels in a crown. This placement accounted for the way the stars retained their fixed positions vis-à-vis one another, each in its own constellation niche, even as the heavens revolved around the Earth every day. While rolling rapidly westward, however, the eighth sphere also betrayed a slow, subtle drift in the opposite direction, which astronomers had long sought to explain. In Copernicus's cosmos, in contrast, the eighth sphere remained stationary. It only *appeared* to move because of the Earth's rotation. But rather than raise this fundamental difference in his critique, Copernicus focused on Werner's technical mistakes.

"In the first place, he went wrong in his calculation of time." Werner had bungled a conversion of Egyptian calendar dates to Julian chronology, so that he assigned certain observations by Ptolemy to the year A.D. 150, when in fact, Copernicus demonstrated, Ptolemy had made those observations eleven years earlier, in 139. Then Werner compounded his initial error by accusing Ptolemy and other ancient astronomers of sloppy observing technique. Here Copernicus lost his temper:

"We must follow in the footsteps of the ancient mathematicians and hold fast to their observations, bequeathed to us like an inheritance. And if anyone on the contrary thinks that the ancients are untrustworthy in this regard, surely the gates of this art are closed to him. Lying before the entrance, he will dream the dreams of the deranged about the motion of the eighth sphere, and will receive his deserts for supposing that he must support his own hallucination by defaming the ancients, who observed all these phenomena with great care and expert skill." In fact, Ptolemy's observations were not as unassailable as Copernicus so passionately insisted, but they were all he had to serve as a basis of comparison, and so he defended them.

Another of his revered ancient Greeks, Hipparchus of Rhodes, had pioneered the exploration of the eighth sphere. More than two hundred years before Ptolemy, around 130 B.C., Hipparchus mapped the positions and comparative brightness of nearly a thousand stars. He intended this work as a baseline for future studies, but tested it first against the few observations that had come down to him from astronomers of prior centuries. Just as he suspected, the constellations showed no change in shape over the course of human memory. However, they had systematically shifted their overall positions. For example, Hipparchus observed the bright star Spica, in the constellation Virgo, at 6° west of the Sun's position at autumnal equinox, while his predecessor Timocharis, on the first night of fall in the fourth century B.C., had seen Spica at 8° west. Every star Hipparchus tested had moved over the interim by the same two degrees—a distance four times the apparent width of the Moon.

To account for this eastward drift of the eighth sphere, Hipparchus's successors posited a ninth: The invisible, external ninth sphere impelled the eighth to turn. But the star-studded eighth, falling just shy of the ninth's pace, lagged behind. The tiny difference between the two went unnoticed night to night, but worked a cumulative effect over decades, amounting to about one degree per century. It would take many ages—many millennia, in fact—for the cycle to come full circle. The extreme slowness and variable rate of the motion, which came to be called the precession of the equinoxes, guaranteed employment for astronomers far into the future. By the time of Werner and Copernicus, accounts of precession had introduced tenth and even eleventh spheres to fine-tune the orientation of the fixed stars.

In modern terms, precession results from the Earth's daily rotation, which produces a planet-wide bulge at the equator. The Sun pulls preferentially on the bulging part, causing the Earth's axis to gyrate slowly over time. It takes twenty-six thousand years for the axis to trace its lazy circle in the sky, at the pace of one degree every seventy-two years. The north pole of the Earth's axis currently points to a star in Ursa Minor (the Little Bear, or Little Dipper) called Polaris, or the Pole Star, also known as the North Star. Precession has brought us to this point, and also warrants that in the next millennium, a different star—Alrai, in the constellation Cepheus, the King—will take over the title of pole star.

Continuing his *Letter*, Copernicus corrected "a second error no less important than the first" and untangled a third "childish blunder" before dismissing Werner.

"What finally is my own opinion concerning the motion of the sphere of the fixed stars?" he asked rhetorically. "Since I intend to set forth my views elsewhere, I have thought it unnecessary and improper to extend this communication further." He wished his friend Bernard sound health and good fortune, and signed off.

In the sphere of sacred affairs, the new Bishop of Rome, Pope Clement VII, watched with horror as Martin Luther's heresy spilled over the German borders into surrounding countries. In Poland, Bishop Ferber could attest that Prussia teemed with Lutherans—not new immigrants but resident Catholics, recently converted. In the diocese of Samland, just east of Varmia in the knights' territory, Bishop Georg von Polenz had renounced his holy vows, publicly endorsed Luther's teachings, and excoriated the Catholic Church. On Easter Sunday 1524, he led a Lutheran Mass in his cathedral. Following this lead, the Bishop of Pomesania, Varmia's western neighbor, also forswore Catholicism for the new evangelism.

Despite Bishop Ferber's angry denunciation of the "heretical wave," certain canons in Varmia viewed the Reform developments differently. Tiedemann Giese responded to Bishop von Polenz's action with a plea for moderation. Giese held that Christians of every stripe should work together to glorify Christ, not denounce one another. "While referring constantly to the spirit of

God," he said of the interfaith dispute, "we are estranging ourselves entirely from love."

Copernicus read Giese's letter and supported its sentiment so strongly that he convinced Giese to publish the content as a pamphlet. Printed in Krakow in February 1525, the *Antilogikon* advocated both tolerance toward Lutherans and defense of Catholic tradition. "Undeniably many things in the Church come close to superstition, and abuses have slipped in," Giese conceded. "But the time of the harvest should be awaited, so that we do not destroy the wheat in order to eradicate the weeds." Eager for his plea to fall on fertile soil, he sent the little book to Wittenberg, to the attention of the scholar and reformer Philip Melanchthon, Luther's closest confederate.

Was the spread of Lutheranism the floodtide presaged by the Great Conjunction of 1524? Astrologers who thought so scrutinized the time and place of Luther's birth to make the case, but even Luther could not say for sure in what year he had been born, whether 1483 or 1484. One practitioner configured Luther's natal chart for October 22, 1484, during a Great Conjunction in the sign of Scorpio, at an hour that conjoined the planets in the ninth house, which was considered the mansion of religion. That fit together nicely, until Luther's brother and mother confirmed an earlier date, November 10, 1483.

Soon another candidate for the destruction prophecy's fulfillment reared up in the form of the Peasant Rebellion, which shed the blood of thousands all over Germany between 1524 and 1525. Ragged peasant militias attacked nobles and armored knights for such things as the right to fish in favorite streams or hunt in forests deemed the private property of the upper class. Although their fight did not concern religion, those peasants who had heard of Martin Luther expected him to support their cause. He disappointed them by failing to intervene: No war of flesh and blood could deter him from his campaign against spiritual wickedness. Soon he turned against them, saying that a peasant in open rebellion stood outside the law of God.

In Varmia, peace finally ended the years of strife between prelates and knights in the spring of 1525, when Albrecht knelt in homage before King Sigismund at Krakow. On April 10, Albrecht ceded his territory to the kingdom, and Sigismund gave it back to him in a new guise, as the Duchy of Prussia, Albrecht's own hereditary fief. As the first Duke of Prussia, he relinquished the title of grand master and, upon his return to Königsberg in May, dissolved the Order of Teutonic Knights. He not only converted, as he had promised Luther he would, but also made Ducal Prussia a fully Protestant state—a new entity on the map of Europe.

Sigismund's agreement with Albrecht stipulated that the duchy would issue no new coins for at least one year, at which time the king would convene an official session to coordinate the ducal currency with the royal. Copernicus dutifully updated his treatise on coinage one last time, in Latin, to accommodate the creation of Ducal Prussia. He added so many specific references to the minting abuses of past grand masters and so many prescriptions for remedy that the essay doubled in length. "What kind of money it will become hereafter, and what its condition is now," Copernicus regretted, "it is shameful and painful to say."

Although Albrecht, as a Catholic, had been the scourge of Varmia, Duke Albrecht the Lutheran behaved more cordially in his dealings with the chapter. He faced different enemies now, among the knights and nobles he had betrayed. Although some members of the old order converted with him, most moved to Germany, where they plotted against him, slandered him, and summoned him to appear before the Holy Roman Emperor's court of justice. When he refused, they proscribed him. Meanwhile Albrecht devoted his energies to the development of his new homeland. He set about establishing schools in every town of Ducal Prussia. On February 12, 1526, he married Princess Dorothea of Denmark, then sired six children in rapid succession.

The King of Poland, who endured the presence of large commu-

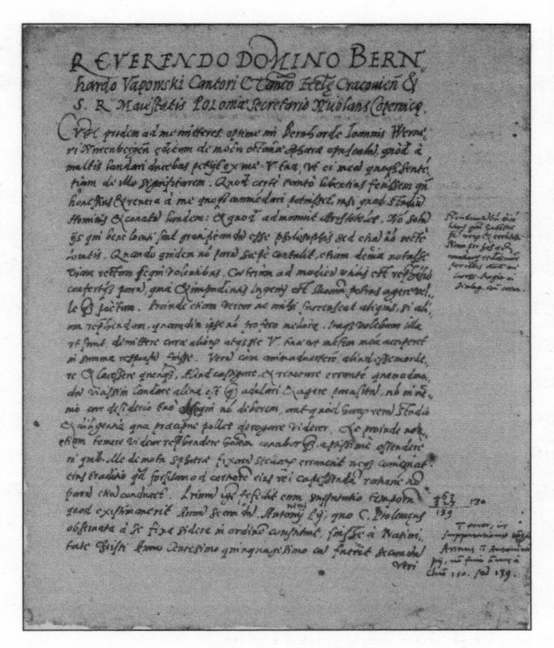
REVERENDO DOMINO BERN
hardo Vapowski Cantori & Canonico Eccl[e] Cracovien[sis] &
S. R. Maiestatis Poloniae Secretario Nicolaus Copernicus.

Like the Brief Sketch, *Copernicus's* Letter Against Werner *was repeatedly hand-copied and forwarded from one interested astronomer to another. This particular copy dates from 1569 and covers ten pages.*

nities of Jews driven into his realm by intolerance in other countries, gradually hardened his stance toward Lutherans. In the spring of 1526, Sigismund ordered the homes of the few Protestants in Krakow to be set on fire. There followed a Lutheran uprising that summer in the Varmian city of Braunsberg. A majority of the citizens had converted to the new religion, including Mayor Philip Teschner, the bastard son of former bishop Lukasz Watzenrode. After the king's forces suppressed the unrest, Bishop Ferber banished all non-Catholics from the diocese. He gave them one month from the date of his edict, September 22, to pack up and go—and on their way elsewhere he insisted they surrender their pro-Lutheran books for burning. Luther himself had published several books and hundreds of pamphlets by this time, as well as his German translation of the New Testament.

Copernicus and Giese could not defend their tolerant attitudes against the objections of the majority of the canons, the bishop, the king, and the pope. They were compelled to vote in favor of the edict that expelled the Lutherans from Varmia. Nevertheless, Giese continued to write in defense of love and acceptance as the Reformation gained ever more new adherents. Several letters allude to these later tracts by Giese, but unfortunately, none was printed or preserved even in manuscript. In one of them, he endeavored to demonstrate the compatibility of Copernicus's theory with the Bible.

One can only guess how Giese squared the Psalmist's praise for a fixed Earth, "immovable and firm," with his friend's contention that the Earth rotated and revolved. Or how he accounted for Joshua's commandment that the Sun stand still, given that Copernicus's Sun never moved. Perhaps Giese did not address each scriptural

King Sigismund Augustus I, shown here in a miniature by Lucas Cranach the Elder.

reference individually, but focused on depicting Copernicus as a godly man, divinely inspired, and blessed with insight into the true workings of the universe. However he constructed his defense, he must have sensed that defense was necessary.

Copernicus also became defensive. By his own admission, he did not feel so confident of his own work as to care nothing for others' opinions of it. He agreed with Giese that readers ignorant of astronomy might easily attack his ideas by twisting chapter and verse to their purposes. Others might laugh at the absurdity of the basic premise. A joke at his expense held that he mistook the Earth for a side of beef, so he put it on a spit and roasted it in the Sun's fire.

The difficulty of the work, coupled with the press of his other duties, may have held Copernicus's anxieties at bay while he was writing his book about the heavenly revolutions. Since he kept no log of his progress, and no one witnessed his isolated toil, it is difficult to say which sections he wrote when, or how long he labored over each. The last observation mentioned in its pages is the one he made on March 12, 1529, when the Moon passed in front of the planet Venus and hid it from view.

"I saw Venus beginning to be occulted by the Moon's dark side midway between both horns at one hour after sunset—the start of the eighth hour after noon," he reported. "This occultation lasted until the end of that hour or a little longer, when I observed the planet emerging westward on the other side. Therefore, at or about the middle of this hour, clearly there was a central conjunction of the Moon and Venus, a spectacle that I witnessed at Frauenburg." He used that observation, paired with others from antiquity, to describe the revolutions and mean motions of Venus, in several pages' worth of diagrams and geometrical proofs spelled out in his small, neat handwriting.*

* Against all odds, the entire handwritten, original manuscript of *On the Revolutions* survives to this day—a bound stack of yellowed paper two hundred sheets thick—in ultrasafe keeping at the Library of the Jagiellonian University in Krakow.

With his book virtually complete by 1535, Copernicus lost courage. He worried that his labored calculations and tables would not yield the perfect match with planetary positions that he had aimed to achieve. He feared the public reaction. He empathized with the ancient sage Pythagoras, who had communicated his most beautiful ideas only to kinsmen and friends, and only by word of mouth, never in writing.

Despite the decades of effort invested in the text, Copernicus eschewed publication. If his theory appeared in print, he said, he would be laughed off the stage. No argument from Giese could change his mind.

Other supporters also tried to sway him. In the summer of 1533, for example, the distinguished linguist and diplomat Johann Albrecht Widmanstetter, then secretary to Pope Clement VII, delivered a lecture on Copernicus's astronomy in the Vatican gardens. Widmanstetter went on, after Clement's death the following year, to serve Nicholas Schönberg, the Cardinal of Capua, and to awaken in him a profound desire to see Copernicus's book published.

On November 1, 1536, the cardinal wrote from Rome: "Some years ago word reached me concerning your proficiency, of which everybody constantly spoke." Cardinal Schönberg had traveled to Poland in 1518 on a peacekeeping mission. Although Albrecht and the Teutonic Order rebuffed his gestures, Bishop Fabian Luzjanski had entertained him in Varmia. "At that time I began to have a very high regard for you, and also to congratulate our contemporaries among whom you enjoyed such great prestige. For I had learned that you had not merely mastered the discoveries of the ancient astronomers uncommonly well, but had also formulated a new cosmology. In it you maintain that the Earth moves; that the Sun occupies the lowest, and thus the central, place in the universe; that the eighth heaven remains perpetually motionless and fixed; and that the Earth, together with the elements included in its sphere, and the Moon, situated between the heavens of Mars and Venus, revolves around the Sun in the period of a year. I have also learned that you have written an exposition of this whole system of astronomy, and have computed the planetary motions and set them down in tables, to the greatest admiration of all. Therefore with the utmost earnestness I entreat you, most learned Sir, unless I inconvenience you, to communicate this discovery of yours to scholars, together with the tables and whatever else you have that is relevant to this subject. Moreover, I have instructed Theodoric of Reden* to have everything copied in your quarters at my expense and dispatched to me. If you gratify my desire in this matter, you will see that you are dealing with a man who is zealous for your reputation and eager to do justice to so fine a talent. Farewell."

Copernicus received this letter, read it perhaps several times over, and then, without responding, filed it away for future use.

* Theodoric of Reden, Copernicus's fellow canon in Varmia, then served as the chapter's representative to the papal court at Rome.

Chapter 6
The Bread Tariff

From one sack of either grain, wheat or rye, cleansed of any grass or weeds before grinding in order that the bread may come out cleaner and purer . . . a careful weighing shows that at least 66 pounds of bread are produced, not counting the weight of the baskets.

—From Copernicus's *Bread Tariff*, ca. 1531

Having banished the Lutherans from Varmia, Bishop Ferber set about putting his own house in order. How shameful, he complained in February of 1531, that in the entire chapter there was barely one priest entitled to celebrate Mass. A general lack of Holy Orders had long characterized the canons at Frauenburg, but now the bishop, who was himself ordained, implored them all to receive their orders—the special grace, imparted by the laying on of hands, that empowered them to administer the sacraments—before Easter. He peered into the private corners of the canons' lives and did not hesitate to censure the slightest infraction. The fifty miles separating the bishop from his subordinates apparently posed no impediment to his learning of their lapses, as when he found out how Canon Copernicus's former housekeeper, though long since discharged, had recently returned to Frauenburg and spent the night with him.

The bishop discreetly took pen in his own hand on this occasion, circumventing his personal secretary. As he recalled—or more likely was reminded by an informant—the housekeeper had married hurriedly after being let go, as though to cover an inconvenient pregnancy, but later separated from her husband. The bishop understood such things. He, too, had been in love as a youth, until his intended jilted him. He fought hard to defend his claim to her hand, producing articles of her clothing as proof of their intimacy, filing suit against her family in the papal court, and traveling to Rome to plead the case himself. When his beloved ultimately wed another man, Maurycy Ferber entered the priesthood.

Now Bishop Ferber called Copernicus to account for his assignation. With the Mother Church under Lutheran attack, no breach of decorum could be borne.

"My noble lord," Copernicus replied on July 27, 1531, "Most Reverend Father in Christ, my gracious and most honorable lord:

"With due expression of respect and deference, I have received your letter. Again you have deigned to write to me with your own hand, conveying an admonition at the outset. In this regard I most humbly ask your Most Reverend Lordship not to overlook the fact that the woman about whom you write to me was given in marriage through no plan or action of mine. But this is what happened. Considering that she had once been my faithful servant, with all my energy and zeal I endeavored to persuade them to remain with each other as respectable spouses. I would venture to call on God as my witness in this matter, and they would both admit it if they were interrogated. But she complained that her husband was impotent, a condition which he acknowledged in court as well as outside. Hence my efforts were in vain. For they argued the case before his Lordship the Dean, your Very Reverend Lordship's nephew, of blessed memory, and then before the Venerable Lord Custodian [Tiedemann Giese]. Hence I cannot say whether their separation came about through him or her or both by mutual consent.

"However, with reference to the matter, I will admit to your Lordship that when she was recently passing through here from the Königsberg fair with the woman from Elbing who employs her, she remained in my house until the next day. But since I realize the bad opinion of me arising therefrom, I shall so order my affairs that nobody will have any proper pretext to suspect evil of me hereafter, especially on account of your Most Reverend Lordship's admonition and exhortation. I want to obey you gladly in all matters, and I should obey you, out of a desire that my services may always be acceptable."

Copernicus's official services at this date made him a guardian of the chapter's counting table. He and Giese, the other guardian, worked side by side collecting installment payments from wealthy individuals who had bought land or commercial operations from the diocese. They also managed the chapter's endowment funds and invested its money in various capital ventures, including the construction of a tavern and inn in Frauenburg. The guardians made new purchases—of properties and sometimes supplies as well, such as "cannon, handguns, lead, and powder for the defense of the Cathedral"—and they paid the salary for the cathedral's preacher.

Probably in the line of his duties as guardian in 1531, Copernicus created his undated *Bread Tariff*. This handwritten document aimed to fix the price of a peasant's daily loaf at an affordable one penny—while simultaneously protecting the interests of the chapter-operated bakery. He advised adjusting the weight of a penny loaf according to the price of grain. Approaching the problem with his typical thoroughness, he calculated a sliding scale to cover a wide range of foreseeable market fluctuations. When harvests were plentiful and grain sold for a pittance, say nine shillings (fifty-four pence) a sack, the peasants would share the bounty, because their penny loaves would weigh in excess of one pound apiece. In lean years, however, should grain prices skyrocket as high as sixty-six shillings per sack, then Copernicus prescribed a loaf of bread had necessarily to shrink to one sixth of a pound. The chapter could recoup its other costs of production—baker's wages, yeast—from the separate sale of bran and chaff.

While guardian, Copernicus made frequent trips to Heilsberg in his capacity as chapter physician, for Bishop Ferber was not a well man. Doctor Copernicus consulted several times with the king's physicians about his treatment. The bishop, partially disabled by an illness years earlier, seemed unable to regain his strength. He perforce sent proxies to meetings where his presence was required. By 1532, age sixty-one, he faced his failing health by naming a coadjutor to aid and eventually succeed him. Not surprisingly, he chose Giese. But King Sigismund interceded and gave the post instead to a career diplomat named Johannes von Höfens, often called Johannes Flaxbinder because his ancestors worked as rope makers, and who signed his poems and letters Johannes Dantiscus, in homage to Danzig, his birthplace.

Dantiscus had spent years angling for a lucrative Varmia canonry. The king first nominated him in 1514, to be Andreas's coadjutor, but the papal court endorsed someone else. The sudden decease of another canon in 1515 again enabled the king to name Dantiscus as a replacement, but this time both the chapter and the pope opposed him. Still the king continued to favor Dantiscus, not forgetting the long love poem he had written three years earlier—the epithalamion for the royal wedding of Sigismund and Barbara. (By 1515, the young queen already lay in her grave, only twenty years old when she died giving birth to her second daughter.)

Dantiscus made his next attempt at a canonry in 1528, while representing Poland at the Spanish court—shortly after he successfully negotiated for Sigismund's new queen, Bona Sforza, the title she desired as Duchess of Bari. Dantiscus's third bid actually won the chapter's support, but failed again in Rome. Finally, on his fourth try, in 1529, Dantiscus got his wish. And then, the very next year, as though to reward his persistence, the king chose Canon Dantiscus to be Bishop of Kulm. The nearby diocese of Kulm was home to the Cistercian convent where Copernicus's sister Barbara had become a nun. Although Kulm could not compare with Varmia in political power or wealth, Dantiscus gladly anticipated putting on the miter.

Johannes Dantiscus,
Bishop of Varmia.

He remained in Spain the next two years, however, to complete his government mission there and also to see to the business of receiving Holy Orders, in order to become a priest prior to ascending to the bishopric.

Returning to Poland in 1532 as Bishop-designate of Kulm, he made a strong candidate for coadjutor to the ailing Bishop Ferber. While he awaited the outcome of that contest, he invited two of his fellow Varmia canons—Copernicus and Felix Reich, but not Giese—to attend his long-overdue investiture ceremony in Löbau on Sunday, April 20, 1533. Both the invited canons begged off.

"Your Lordship, Most Reverend Father in Christ," Copernicus wrote, "I have received your Most Reverend Lordship's letter, from which I quite understand your kindliness, graciousness, and good will toward me. . . . This is surely attributable, not to my services but rather, in my opinion, to your Most Reverend Lordship's well known generosity. Would that I might some day chance to be so deserving! I am of course more delighted than it is in my power to say that I have found such a patron and protector.

"Your Most Reverend Lordship, however, requests that I join

you on the 20th of this month. Although I would do so with the greatest pleasure, since I have no insignificant reason for attending so eminent a friend and patron, yet the misfortune has befallen me that at that very time Canon Felix and I are required by certain business and by compelling reasons to remain at our stations. I therefore ask your Most Reverend Lordship to excuse my absence at that time. At any other time I am unreservedly ready, as I should be, to call on your Most Reverend Lordship and do whatever pleases you, to whom I am indebted in very many other ways, provided that your Most Reverend Lordship so indicates to me at some other time. I acknowledge that hereafter I should not so much satisfy your requests as execute your commands."

Yet he continued to avoid contact. Two years later, in a letter declining an invitation from Dantiscus to a family wedding, Copernicus apologized for still not having called at Löbau. Now he regretted that some new ineluctable duty in Varmia would keep him from the nuptials. "Therefore deign to excuse my personal absence and preserve your old attitude toward me, even though I shall not be present. For, a meeting of the minds usually counts for even more than a meeting of the bodies."

Copernicus could not distance himself indefinitely from Dantiscus, however. On July 1, 1537, Bishop Ferber died of a stroke while Doctor Nicolaus was rushing to his bedside. Soon after, as had been predetermined, the canons elected Dantiscus the Bishop of Varmia. As for the newly vacant bishop's seat in Kulm, that was to be filled by Tiedemann Giese. He would need to move to the much poorer diocese of Kulm, southwest of Varmia, away from Copernicus and his other accustomed associates, though still retaining his position as their fellow canon.

Whatever resentment Dantiscus had stored up during the years of snubbing by the Varmia canons now erupted in a show of vindictive behavior. He began with Giese. The coadjutor contest with that man had been a little too close, and Dantiscus determined to deny Giese the possibility of becoming Bishop of Varmia—ever—even in the event of Dantiscus's death. Although Giese, five years older than Dantiscus, seemed unlikely to outlive him, still Dantiscus sought to squelch any chance. To that end, he demanded that Giese, as Bishop of Kulm, give up his canonry in Varmia. And he pressured Copernicus to push Giese's resignation through the chapter's official channels.

"With regard to that matter of the canonries, which your Most Reverend Lordship entrusted to me," Copernicus wrote from Frauenburg on Easter Sunday 1538 to Dantiscus in Heilsberg, "I have received the plan and shared it with the Most Reverend Bishop of Kulm." But, Copernicus regretted, the matter could not be referred to the chapter at this moment, since other more pressing cases had to be heard first. "When that is done, there will be a better opportunity to bring up the proposal, unless another plan is conceived in the meantime by your Most Reverend Lordship, to whom I wish my services to be acceptable."

A surprisingly rich cache of correspondence from this period documents the intrigues that enveloped the cathedral. One letter, from Queen Bona, congratulated Dantiscus on his rise to the bishopric of Varmia and reminded him of his new duty to provide good German horses, "even from his own stable," to her son, Prince Sigismund Augustus. Queen Bona further urged Dantiscus to relinquish his canonry in Varmia, on the grounds that "it is unbecoming to be the bishop and a canon in one and the same church." Whether any previous bishop had ever made such a concession before, Dantiscus had little to lose by acquiescing on this score. As bishop, his landholdings and income dwarfed the annual receipts from his canonry. He readily let it go—not to Bona's handpicked worthy (a second cousin of Copernicus), but to his own protégé Stanislaw Hozjusz. This move gave Dantiscus a new ally in the chapter. It also allowed him to wield Bona's concerns about being bishop and canon in the same church as a weapon against Giese. In point of fact, Giese was bishop of one church (Kulm) and canon of a different one (Varmia), but Dantiscus waved aside that detail.

While Copernicus stalled to protect Giese, Dantiscus shifted his focus to the issue of the canons' female housekeepers. Unless these were relatives, the bishop proclaimed, they must be dismissed. Copernicus's present cook, Anna Schilling, a married woman separated from her legal husband, may have been the same individual who had aroused the late Bishop Ferber's outrage.

"My lord, Most Reverend Father in Christ, most gracious lord, to be heeded by me in everything," Copernicus opened his reply to Dantiscus on December 2, 1538, "I acknowledge your Most Reverend Lordship's quite fatherly, and more than fatherly admonition, which I have felt even in my innermost being. I have not in the least forgotten the earlier one, which your Most Reverend Lordship delivered in person and in general. Although I wanted to do what you advised, nevertheless it was not easy to find a proper female relative forthwith, and therefore I intended to terminate this matter by the time of the Easter holidays. Now, however, lest your Most Reverend Lordship suppose that I am looking for an excuse to procrastinate, I have shortened the period to a month, that is, to the Christmas holidays, since it could not be shorter, as your Most Reverend Lordship may realize. For as far as I can, I want to avoid offending all good people, and still less your Most Reverend Lordship. To you, who have deserved my reverence, respect, and affection in the highest degree, I devote myself with all my faculties."

In a Greek flourish at the end of this letter, instead of identifying his locale as "Frauenburg," in the usual way, Copernicus wrote "Gynopolis." One could argue that "Gynopolis" matched the literal meaning of the place name. But, given the context of the discussion, Copernicus seems mischievously to suggest that "the city of Our Lady" has gained new significance for its less exalted women—the female housekeepers.

"I have now done what I should not or could not in any way have failed to do," he capitulated in early January 1539. "I hope that what I have done in this matter quite accords with your Most Reverend Lordship's warnings." She had gone. Her leaving, however, did not satisfy Dantiscus's urge to punish Copernicus and two other canons who had flouted him in the matter of the housekeepers. The bishop now entered into clandestine communication with Canon Felix Reich, who had no compunctions about ratting on his brethren—even as Copernicus was treating him for a bleeding ulcer. In exchange for inside information, Bishop Dantiscus volunteered to procure whatever Reich requested against his painful condition, including light beer in wholesome daily doses and unlimited quantities of Hungarian wine to strengthen his heart.

Reich's years as a notary had made him an expert on legal protocol. Now he instructed Bishop Dantiscus to send sealed writs addressed individually to the three scandalous canons. Separate writs for the women should be sent to the local priest, who would issue the warnings to them.

"Care should also be taken to omit from the letters to the other two, who do not have legal husbands," Reich stipulated, "what is in that earlier letter concerning Nicolaus's cook, who does have a legal husband. The impending commencement of the proceeding against the women too will strike terror to no small degree.

"Whatever the situation may be, may Your Most Reverend Lordship act firmly. God Almighty will strengthen your arm so that you may conduct to a happy ending what you initiated out of zeal. As much as we can, all of us will help make a success of this affair. However, your Most Reverend Lordship must take care nevertheless in commencing the proceeding with the force of law not to introduce in your future letters anything contrary to formal and customary legal style, as it is called. For it often happens that even the tiniest clause may spoil an entire case, so that it is declared null and void if it comes before a higher judge."

Dantiscus's own dalliance in Spain had produced at least one illegitimate daughter and a son who died in infancy, but this did not deter him from prosecuting the canons and their concubines. He tried to proceed as Reich suggested, but could not keep the paperwork free of errors.

"I am sending back all the letters because in one a serious scribal error must be corrected, and that cannot be done here," Reich complained on January 23. The bishop's scribe had addressed one of the letters to Heinrich Scultetus, when in truth that canon's first name was Alexander—the acknowledged father of several children with his live-in maidservant.

"Moreover, in my previous letter I warned about the banishment of 'ten miles' and 'outside the diocese,' since your Most Reverend Lordship does not have the power to banish anyone beyond your own diocese, which in certain places (as here in Frauenburg) does not extend farther than one mile. Consequently it would be necessary to delete this reference to ten miles as the distance to which the women are relegated." Once the bishop had seen to these and miscellaneous other corrections, Reich advised, he should return the documents with caution:

"The letters to the canons should be tied up separately, and the letters to the cooks in like manner, and these sealed in one envelope and addressed to the priest. Otherwise a great mishap could occur. For if open letters to the cooks fell into the hands of those three canons, I have no doubt that, being intercepted, the letters could not accomplish their purpose. Your Most Reverend Lordship will therefore instruct your courier on his return to deliver to the priest the documents pertaining to the women first, ahead of everything, and afterwards the documents concerning the canons to any canon. The latter will undoubtedly give each one his own and, if need be, accost each one most circumspectly. Otherwise he will heap no small suspicion on me.

"I suppose your Most Reverend Lordship has a sound reason for writing very briefly to Nicolaus, and it does not matter a great deal. Undoubtedly they all coordinate everything with one another."

Before the bishop could act or reply, Reich wrote again on January 27 to thank Dantiscus for a new delivery of wine and beer, and also to forestall another faux pas.

"Together with the wine from Allenstein, through the effort of the venerable Administrator there, your letter to the Chapter was also delivered to me. I am afraid, however, that it may contain something about the proceedings against the canons' cooks and against the canons themselves. I do not dare deliver the letter to the Chapter lest its members cause a disturbance in this affair. . . . I beg you not to be angry because without any instructions I withheld the letter in accordance with my own judgment."

Reich's meddling soon ended with the aggravation of his illnesses, which led to his death on March 1. The chapter buried him the following day. On March 3, Copernicus claimed Reich's vacant canonry for his relative, Raphael Konopacki, previously nominated by Queen Bona. In April the young man's father—George Konopacki, the governor of Pomerania—wrote to Dantiscus to smooth Raphael's way into the cathedral community: "I most humbly beseech your Most Reverend Paternity (for I have understood that you always support my affairs and those of my son with special kindness, and we look upon you as the anchor of our hope) to deign to assist him with the same utmost kindness and good-will and help."

Right at the same time that Raphael came to Frauenburg to assume his canonry, in May of 1539, another youth—a brilliant mathematician—also entered the city in search of Canon Copernicus. No one had invited him or even suspected his arrival. Had he sent advance notice of his intent to visit, he doubtless would have been advised to stay far away from Varmia. Bishop Dantiscus's most recent anti-heresy pronouncement, issued in March, reiterated the exclusion of all Lutherans from the province—and twenty-five-year-old Georg Joachim Rheticus was not only Lutheran but a professor at Luther's own university in Wittenberg. He had lectured there about a new direction for the ancient art of astrology, which he hoped to establish as a respected science. Ruing mundane abuses of astrology such as selecting a good time for a business transaction, Rheticus believed the stars spoke only of the gravest matters: A horoscope signaled an individual's place in the world and his ultimate fate, not the minutiae of his daily life. If properly understood, heavenly signs would predict the emergence of religious prophets and the rise or fall of secular empires.

Rheticus had left Wittenberg the previous autumn, in October 1538, amidst squabbles over some vulgar verses written by a friend of his to mock Martin Luther. On leave from his teaching duties, he had spent months traveling around Germany, intent on seeking out the best practitioners to enlighten his understanding of astronomy and its astrological applications. In Nuremberg, he learned for the first time of the Polish canon who accounted for the celestial motions by centering them on the Sun. This concept seemed to Rheticus a redemptive cosmology, and without delay he undertook the five-hundred-mile journey to northern Prussia to learn the details of the theory from its source.

The name Rheticus, like that of Dantiscus, derived from a place instead of a person. Rheticus would have used his real surname if allowed to, but his family had been stripped of that privilege after his father, the physician Georg Iserin of Feldkirch, was beheaded for his crimes in 1528. Some accusers branded Iserin a sorcerer; others called him a thief who had come into their homes to give medical care but left with their valuables. Rheticus, fourteen at his father's execution, first took his mother's maiden name, de Porris, but later changed it to the more German-sounding von Lauchen, which meant the same thing: "of the leeks." At the start of his university studies at

Wittenberg in 1532, he acquired the toponym that tied him to his Alpine homeland of Rhaetia.

Rheticus had begun school with an interest in medicine but displayed an uncanny aptitude for numbers. He quickly came under the wing of the renowned humanist scholar Philip Melanchthon, "the teacher of Germany" and also Luther's trusted supervisor of university affairs. As Rheticus later reported, the fatherly Melanchthon pushed him toward the field of mathematics—to the study of arithmetic, geometry, and astronomy. In 1536, rather than lose Rheticus after conferring on him the master of arts degree, Melanchthon established a second mathematics professorship especially for him to fill. Rheticus, now twenty-two, marked his transformation from student to faculty member at Wittenberg with a public inaugural address. Feeling awkward at the center of such attention, he

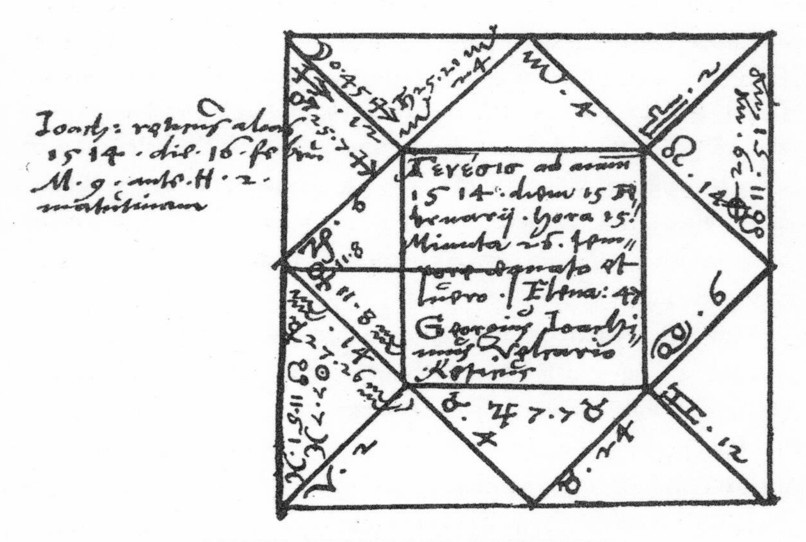

NATAL CHART FOR RHETICUS

The dire prospects suggested in this horoscope for Georg Joachim Rheticus caused his student Nicholas Gugler, who drew the chart, to recalculate the professor's birth time and date. The true date of February 16, written in the margin, disagrees with the more favorable date in the diagram, February 15.

warned his audience that he was "shy by nature," and most dearly cherished "those arts that love hiding-places, and do not earn applause among the crowds." Nevertheless he acquitted himself admirably in his oration. "It is characteristic of the honorable mind," he said, "to love nothing more ardently than truth, and, inspired by this desire, to seek a genuine science of universal nature, of religions, of the movements and effects of the heavens, of the causes of change, not only of animated bodies, but also of cities and realms, of the origins of noble duties and of other such things." Mathematics, he avowed, united all these pursuits.

In addition to their mutual regard, Rheticus and Melanchthon shared a devotion to astrology that did not include the dubious Luther. "Astrology is framed by the devil," scoffed Luther at table one day, "for the star-peepers presage nothing that is good out of the planets." As espoused by Melanchthon, however, astrology contained no taint of devil or magic. Its tenets were upheld by the text of Genesis, which told how God placed "lights in the firmament of the heaven" not only "to divide the day from the night," but also to serve as "signs."

Rheticus had of course cast his own horoscope. His birth, in the very early hours of February 16, 1514, coincided with a conjunction of the Moon and Saturn in the twelfth house. There was no mistaking the ominous import of these conditions: They augured an abnormally short life span. As an adept astrologer, Rheticus knew several ways to rectify a bad chart. In one experiment, he technically escaped his fate by moving his birthday to the previous day, February 15, and changing the time from nine minutes before the second hour of the morning to 3:26 in the afternoon. These alterations divided Saturn from the Moon and moved them into separate houses, granting him a reprieve. But it seems unlikely that such fiddling would have rid Rheticus of the fear of impending doom. A sword, like the one that had severed his father's head, hung menacingly over his own.

Part Two
Interplay

You, who wish to study great and wonderful things, who wonder about the movement of the stars, must read these theorems about triangles. Knowing these ideas will open the door to all of astronomy.

—Johannes Müller, known as Regiomontanus (1436–1476), author of the *Epitome of Ptolemy's Almagest* and *On Triangles*

Then spake Joshua to the Lord in the day when the Lord delivered up the Amorites before the children of Israel, and he said in the sight of Israel, Sun, stand thou still upon Gibeon; and thou, Moon, in the valley of Ajalon.
And the Sun stood still, and the Moon stayed, until the people had avenged themselves upon their enemies. Is not this written in the book of Jasher? So the Sun stood still in the midst of heaven, and hasted not to go down about a whole day.
And there was no day like that before it or after it, that the Lord hearkened unto the voice of a man: for the Lord fought for Israel.

—Joshua 10:12–14

AND THE SUN STOOD STILL

A Play in Two Acts

❧ ❧

CAST OF CHARACTERS

COPERNICUS, age 65, physician and canon (church administrator) in Varmia, northern Poland

BISHOP (of Varmia), age 53

FRANZ, age 14, the BISHOP's acolyte

RHETICUS, age 25, mathematician from Wittenberg

ANNA, age 45, housekeeper to COPERNICUS

GIESE, age 58, Bishop of Kulm (another diocese in northern Poland) and canon of Varmia

A native of Torun, Copernicus lived thirty years in Frauenburg, "the city of Our Lady," in the shadow of its medieval cathedral. Frauenburg, the seat of the Varmia diocese, is the setting for the play.

ACT I

Scene i. In the Bishop's bedroom
HOUSE CALL

The time is May 1539, in northern Poland, near a medieval cathedral ringed by fortified walls.

Darkness. The sound of someone retching. Lights up on Copernicus, *standing over the* Bishop, *his patient, who sits on the edge of the bed in his richly appointed apartment, vomiting into a basin.*

Franz, *the frightened young acolyte, hovers and helps as needed.*

Bishop. Oh, God. Oh, Heaven help me.
Copernicus. I think that was the last of it, Your Reverence.

Copernicus *takes the basin, but the* Bishop *grabs it and vomits one more time, then collapses back onto his bed.*

Bishop. Oh, Lord have mercy. Ohhh.
Copernicus. Take this away, Franz. There's a good lad.

Franz *bows, exits with the basin.*

The Bishop *writhes, groans.*

Bishop. I thought I would surely die.
Copernicus. The pain will subside, now that the emetic has rid your body of that toxin. You should be fine by tomorrow.
Bishop. "Toxin"?!
Copernicus. It's all gone now. You've expelled it.
Bishop. Poison?!
Copernicus. No. No, a toxin is . . .

BISHOP. Lutherans!

COPERNICUS. Now, now.

BISHOP. I've been poisoned. If you hadn't come, I'd be dead.

COPERNICUS. Not poison, Your Reverence. More likely something you ate.

BISHOP. Of course it was something I ate. They put it in my food. How else would they get it into me?

COPERNICUS. It could have been a bite of rotten fish.

BISHOP. The kitchen staff! That shifty-eyed cook must be a Lutheran sympathizer.

COPERNICUS. Just an ordinary bit of bad fish. Not poison.

BISHOP. The Lutherans want to assassinate me.

COPERNICUS. Or maybe too much eel. Your Reverence is extremely fond of eel.

BISHOP. I should have known better. Banishing them from the province was not enough to eliminate the threat.

COPERNICUS. Swallow this, Your Reverence. To settle your nerves and bring on sleep.

BISHOP. Sleep?! How can I sleep when Lutheran dogs are stalking me?

COPERNICUS. Sleep will be the best thing now.

BISHOP. Worse than dogs. Vermin! Evil and dangerous. They simply ignore the law. They're below the law. Here in our midst, waiting for the moment to strike. Oh, Nicholas, what if they try it again?! Suppose they make another attempt on my life, and you don't get here in time? What if . . . ?

COPERNICUS. Take this, please, Your Reverence.

The BISHOP *refuses the medicine, pushes* COPERNICUS *away.*

BISHOP. We must prosecute them more forcefully. Threaten offenders with harsher punishment. I won't let them get me the way they got Bishop Ferber.

COPERNICUS. Bishop Ferber?

BISHOP. I see it all now.

COPERNICUS. No one poisoned Bishop Ferber.

BISHOP. They didn't have to! He let them do as they pleased. They walked all over him. Until God Almighty intervened to smite him for not smiting them.

COPERNICUS. Bishop Ferber died of syphilis.
BISHOP. One of God's favorite punishments.

FRANZ *returns, busies himself tidying the room.*

BISHOP. Aah! It's done now. He's in his grave, and may he rest in peace. But why did he have to leave the whole Lutheran mess in my hands?

The BISHOP *starts to get out of bed, but* COPERNICUS *restrains him.*

BISHOP. I must deal harshly with them. I cannot afford to show weakness.

COPERNICUS *succeeds in settling the* BISHOP *in bed.*

BISHOP. Oh, my heart. Franz! Bring me a glass of my Moldavian wine. And one for Doctor Copernicus.

FRANZ *exits.*

BISHOP. That wine is the better tonic. To strengthen me for the fight. I'll issue a new edict. This time I'll ban their books, too, so they can't . . . Ban them and burn them. And their music is anathema.

FRANZ *returns with two filled glasses.*

BISHOP. No one may sing those hateful hymns any longer. On pain of . . . Aaahhh. Here's our spirit.

COPERNICUS *takes a glass, puts the medicine in it, and hands it to the* BISHOP, *who drinks it.*

BISHOP. Agh! Curse that poison! It's killed the taste of pleasure.
COPERNICUS. (*raising the other glass*) To good health, Your Reverence.
BISHOP. Amen.

They both drink.

COPERNICUS *gives his glass to* FRANZ, *prepares to leave.*

BISHOP. Don't hurry off, Nicholas.
COPERNICUS. Sleep will be the best company now, and will soon arrive.

BISHOP. Stay and have another glass. Your conversation is a comfort to me.

At a signal from the BISHOP, FRANZ *exits.*

COPERNICUS. I should go now.

BISHOP. What's your hurry?

COPERNICUS. I must not keep Your Reverence from the sleep I have prescribed.

BISHOP. Rushing home to your . . . manly duties?

COPERNICUS. My . . . ?

BISHOP. Don't give me that innocent look. You know what I'm talking about.

COPERNICUS. I don't . . .

BISHOP. Your harlot!

COPERNICUS. You mean my . . . ?

BISHOP. You know damn well who I mean.

COPERNICUS. She's not . . .

BISHOP. You should get rid of her.

COPERNICUS. But she . . .

BISHOP. She's not Lutheran, is she?

COPERNICUS. No.

BISHOP. Get rid of her anyway.

FRANZ *returns with more wine, pours.*

BISHOP. I'm serious, Nicholas. I want her out of your house. It looks bad. Keeping an unmarried woman like that.

COPERNICUS. She cooks and cleans for me.

BISHOP. She's not even related to you. It's unseemly.

COPERNICUS. If I had a female relative who could . . .

BISHOP. And much too good-looking.

COPERNICUS. She's done nothing wrong.

BISHOP. Get yourself an old hag. Or a boy, to take care of your . . . needs. (*drinks his second glass*) Listen, Nicholas. For myself, I don't care who's in your bed. I understand a man's appetites. God knows, I sowed my oats. Fathered a child or two, here and there, before . . . But it's different now. With Luther and his devils screaming to high Heaven and Rome about

Church abuses, a man in your position . . . A canon of this cathedral! You must appear above reproach.

COPERNICUS. Yes, Your Reverence.

BISHOP. (*yawning*) Go on home now. Tell her to find a new position. Someplace far away from here.

COPERNICUS *exits.*

Lights fade. A bell tolls the hour: 3 o'clock.

SCENE ii. OUTSIDE COPERNICUS'S HOUSE STRANGER ARRIVES

Minutes later, outside the cathedral wall, COPERNICUS *walks home with a lantern. At the door of his house, he discovers* RHETICUS *lying on the ground.* COPERNICUS *jumps back, then bends down to examine him, checking his pulse, loosening his clothing.*

RHETICUS *awakens with a cry, lashes out.*

RHETICUS. Ho! Get off me!

COPERNICUS. Are you ill?

RHETICUS. Get away from me! Thief!

COPERNICUS. I was just trying to . . .

RHETICUS. Thief!

RHETICUS *pounces on* COPERNICUS*; they scuffle.*

COPERNICUS. No! Oh!

RHETICUS. What did you take?

COPERNICUS. I didn't . . . Oh!

RHETICUS. Give it back!

COPERNICUS. Don't!

RHETICUS. (*pinning* COPERNICUS *to the ground*) Give it back or I'll strangle you.

COPERNICUS. (*choked, gasping*) I'm a doctor.
RHETICUS. What?
COPERNICUS. I'm a doctor. I thought you were hurt. I was trying to help.

RHETICUS *releases* COPERNICUS, *then stands, pats his body to make sure he has his belongings, looks in his satchel.*

COPERNICUS *tries to stand.*

RHETICUS. Don't move.
COPERNICUS. Who are you?
RHETICUS. You scared me to death.
COPERNICUS. I thought you were dead. I thought . . .
RHETICUS. I was just waiting there, when you came along and . . .
COPERNICUS. You were lying on the ground.
RHETICUS. Right there. I was sitting right over there.
COPERNICUS. (*struggling to rise*) Who are you?
RHETICUS. I was waiting to see . . .
COPERNICUS. Ow!
RHETICUS. Are you hurt?
COPERNICUS. My ankle. I think I . . .
RHETICUS. You must have twisted it when you fell.
COPERNICUS. (*indicating* RHETICUS'*s satchel*) Give me that, will you?

COPERNICUS *props the satchel under his foot, ties his handkerchief around his ankle.*

RHETICUS. I'm sorry I hurt you, Doctor. I didn't know . . .
COPERNICUS. What are you doing here?
RHETICUS. I'm waiting for Canon Copernicus. This is his house, isn't it?
COPERNICUS. What do you want with him?
RHETICUS. Is he sick? Is that why you've come?
COPERNICUS. No, he's not sick.
RHETICUS. Thank God. Imagine if I'd come all this way, only to find the great canon, the *starry* canon, too sick to receive me.
COPERNICUS. What did you call him?

RHETICUS. Please forgive me, Doctor. I don't normally get into fistfights. You may not believe this, but I'm a scholar by profession.

COPERNICUS. You?

RHETICUS. A mathematician.

COPERNICUS. Really?

RHETICUS. Professor of mathematics, in fact. (*extending his hand*) My name is Rheticus, sir. Georg Joachim Rheticus.

COPERNICUS *starts to extend his own hand.*

RHETICUS. Of the mathematics faculty at Wittenberg.

COPERNICUS. (*withdrawing his hand*) Wittenberg?!

RHETICUS. You've heard of it, of course?

COPERNICUS. You came here? From Wittenberg?

RHETICUS. To tell you the truth, I was actually stopping at Nuremberg when I decided to come here.

COPERNICUS. But Wittenberg is . . .

RHETICUS. Nuremberg is even farther. It added another hundred miles to my journey.

COPERNICUS. But it's not safe.

RHETICUS. Not safe to travel anywhere these days. Between the bandits and the dogs. And the rain! Twice in one day I was almost drowned fording rivers.

COPERNICUS. From Wittenberg.

RHETICUS. The canon will know its reputation . . .

COPERNICUS. Indeed.

RHETICUS. As a place where the study of mathematics has always flourished.

COPERNICUS. (*returning the satchel*) Here, take this back.

RHETICUS. Keep it, please. Use it as long as you like.

COPERNICUS. (*rising with difficulty*) This is Poland, Professor. Catholic Poland.

RHETICUS. I'm sure Canon Copernicus will welcome me, as a natural philosopher.

COPERNICUS. He will do no such thing. He cannot.

RHETICUS. We'll see what he . . . Whoa, there, Doctor. Are you sure you can walk?

COPERNICUS. (*indicating the house*) I don't have far to go.

RHETICUS. Here? But . . . You mean you are . . . ?

COPERNICUS *nods.*

RHETICUS. (*kneeling*) Oh, no. Oh, my God! Oh, please forgive me!

COPERNICUS. Now, now. Don't . . .

RHETICUS. All the times I pictured our meeting, and to think . . . Dear Lord, how I've botched things!

COPERNICUS. It's all right. I'm fine. But you had better move on. This is no place for you.

RHETICUS. If only you knew how I . . .

COPERNICUS. Please, get up.

RHETICUS. The whole way here, I rehearsed, over and over, what I would say when I met you.

COPERNICUS. Say it, then. On your feet.

RHETICUS. (*rising*) Canon Copernicus, I . . . Is that the right way to address you, sir? Or should I call you "Father"? Did you say you were a doctor?

COPERNICUS. It doesn't matter. Say your piece.

RHETICUS. Begging your pardon, Canon sir. Doctor. I have letters here from . . . (*fishing in his satchel*) Letters of introduction from . . .

COPERNICUS. Don't bother with that.

RHETICUS. Here they are, sir. From Schöner in Nuremberg. And another one here from Hartmann, and also Peter Apian, and . . .

COPERNICUS. Did you say, from Schöner?

RHETICUS. Yes, sir. (*handing him the letter*) Here, see for yourself. He was gracious enough to let me stay several weeks with him, in his home. This one is from Camerarius, in Tubingen. He tried to convince me not to look for you. He said you must be dead by now. Excuse me, sir. I meant no offense. It's just that no one has heard from you in so long. They're all waiting. They wonder why you've kept silent all this time.

COPERNICUS *finishes reading the letter.*

COPERNICUS. I have . . . nothing to say.

RHETICUS. You are too modest, sir. What you've done . . . Why, you have made the greatest leap in astronomy since . . . Ptolemy introduced the equant. (*brandishing the letters*) Everyone speaks of you. "The Polish canon," they call you, "who spins the Earth and makes the Sun and stars stand still." They say you've been working at your thesis for more years than I've lived.

COPERNICUS. I'm finished with all that now.

RHETICUS. You've finished? You're ready to release the details?

COPERNICUS. There's nothing here for you, Professor. You should go back to Wittenberg. I'm sure your students miss you.

RHETICUS. With all due respect, sir. Classes are suspended for the summer holiday. And besides, I have been on special leave the past two semesters, on a personal mission, meeting with the most learned mathematicians of our time. I believe that you, sir, are the culmination of my quest. The very key to the perfection of the heavenly spheres.

Beat.

RHETICUS. Sir, I seek to restore the queen of mathematics, that is, Astronomy, to her palace, as she deserves, and to redraw the boundaries of her kingdom.

COPERNICUS. I can't help you.

RHETICUS. Only you can help me.

COPERNICUS. Good night, Professor.

COPERNICUS *starts for his door.*

RHETICUS. And I can help you, too. Let me tell you my plan.

COPERNICUS. I wish you a safe journey.

RHETICUS. Hear me out!

COPERNICUS. Hush. I'm telling you, for your own safety, to leave this place.

RHETICUS. In the middle of the night? After I've traveled weeks just to find you?

Blackout.

Scene iii. Inside Copernicus's house
GIFTS

Moments later, COPERNICUS *enters the main room of the house (not as ornate as the* BISHOP'S *palace, though comfortable).*

The house is dimly lit, but very gradually, through the scene, dawn begins to lighten the room.

RHETICUS *follows* COPERNICUS *into the room.*

COPERNICUS. You can spread out your bedding in there.

RHETICUS. I'm not the least bit tired. I wonder whether we could just take a few moments to . . .

COPERNICUS. In the pantry, you'll find some bread. You may take what you need for your . . .

RHETICUS. Oh! I almost forgot! In all the confusion, sir, I never gave you the gifts I brought.

COPERNICUS. I can't accept gifts.

RHETICUS. (*pulling books from his satchel*) You must. This is Ptolemy.

COPERNICUS. Thank you, but no. I have studied Ptolemy, of course. Every astronomer has studied . . .

RHETICUS. You would have read a Latin translation, from the Arabic or Hebrew. This is the original Greek text.

COPERNICUS. Oh?

RHETICUS. Only recently recovered and now published for the first time.

COPERNICUS. Let me just have a look at that.

RHETICUS. And this is Euclid's *Geometry*, also in Greek. And here, Regiomontanus, on triangles. I love the part at the beginning, where he says, "No one can bypass the science of triangles and reach a satisfying knowledge of the stars."

COPERNICUS. These are magnificent volumes.

RHETICUS. I chose the ones I knew you would like, Canon sir.

COPERNICUS. I couldn't. I . . . You keep them for your own library.

RHETICUS. I've already inscribed them to you.

COPERNICUS. (*reading*) "To N. Copernicus, my teacher . . ." Your teacher?

RHETICUS. I was hoping . . .

COPERNICUS. I have no students.

RHETICUS. I know that, sir.

COPERNICUS. No followers of any kind.

RHETICUS. That is why I have come.

COPERNICUS. I'm sorry, Professor.

RHETICUS. To be your disciple. Whatever problems have interfered, kept you from bringing your work to completion, I want to help you solve them. I showed you my letters. Even Melanchthon says I have exceptional aptitude.

COPERNICUS. Philip Melanchthon?!

RHETICUS. "The teacher of Germany," yes.

COPERNICUS. Luther's own chosen successor? His right hand?!

RHETICUS. He said I was born to study mathematics.

COPERNICUS. Tell me, Professor: Are you on intimate terms with Luther, too?

RHETICUS. Oh, now I see what you . . . But I swear to you, sir, I do not share the Reverend Luther's opinion of your ideas. No, indeed.

COPERNICUS. Martin Luther has an opinion about my . . . ?

RHETICUS. It's only his opinion. Whereas, I feel, astronomy requires precisely the kind of bold new approach that you take.

COPERNICUS. What does he say about it?

RHETICUS. Oh. Things come up at faculty meetings. Lunches. At table. You know how it is.

COPERNICUS. No.

RHETICUS. Someone gave him the gist of it, and . . .

COPERNICUS. And?

RHETICUS. Well . . .

COPERNICUS. What did he say?

RHETICUS. He said, only a fool would turn the whole of astronomy upside down, merely for the sake of novelty.

Beat.

COPERNICUS. I suppose "fool" is a mild insult, coming from him.

RHETICUS. And of course he knows nothing of mathematics. He only rejected your theory because it contradicts the Bible. He quoted Joshua 10:12. You know the part, where Joshua says, "Sun, stand thou still upon Gibeon."

COPERNICUS. Yes, yes. I know it all too well.

RHETICUS. "And thou, Moon, in the valley of Ajalon."

COPERNICUS & RHETICUS. (*together*) "And the Sun stood still."

RHETICUS. Exactly, sir. The Sun stood still. And that's his point. Because, if the Sun were already standing still, as you claim, then why would Joshua have commanded it to do so?

COPERNICUS. Why do you think?

RHETICUS. I say, mathematics is for mathematicians. Scripture doesn't enter into it.

COPERNICUS. Is that what you told him?

ANNA *enters the room, in robe, shawl, and nightcap, carrying a candle.*

ANNA. Is everything all right?

COPERNICUS. (*going to her*) Anna. What are you doing up at this hour?

ANNA. Have you been hurt, Mikoj?

RHETICUS. She wouldn't let me in when I arrived.

COPERNICUS. It's nothing.

ANNA. What happened?

COPERNICUS. I sprained it, that's all.

RHETICUS. I thought she was going to throw water on me.

ANNA. He came to the door after you left.

RHETICUS. I tried to tell her I was only . . .

ANNA. I was afraid to let him in.

COPERNICUS. You did the right thing.

While ANNA *and* COPERNICUS *talk,* RHETICUS *fishes in his satchel for some papers.*

ANNA. What does he want?

COPERNICUS. He? He'll be leaving in just a few . . .

ANNA. What about the bishop?

COPERNICUS. Oh!

ANNA. Not . . . ?

COPERNICUS. No, no. He's . . . resting. Go back to bed, dear. I'm sorry we woke you.

ANNA. I've been waiting up worrying the whole night, Mikoj.

COPERNICUS. Everything's all right now.

ANNA. I don't like the look of him.

COPERNICUS. I'll take care of this. Don't worry.

Reluctantly, ANNA *exits.*

COPERNICUS. I hate to be inhospitable, Professor. I wish . . .

RHETICUS. There's something else I brought you, sir.

COPERNICUS. No, please. No more gifts.

RHETICUS. These are from Schöner. Some recent observations he collected, of Mercury. He insisted that I give them to you.

COPERNICUS *takes the sheets of paper, studies them.*

RHETICUS. He didn't make the observations himself. He said he got them from someone else, but he remembered that you always wanted . . .

COPERNICUS *shakes his head with wonder, nods in admiration, sighs.*

RHETICUS. He said you'd be pleased. He was sure you would. You haven't really given it up, have you, sir? You must still be working on it. Am I right? Sir?

COPERNICUS. Hm?

RHETICUS. I said, you haven't quit. Have you? It's just taking time. Isn't that right? That's why I thought I could . . .

COPERNICUS. No. I'm sorry. Even if I wanted to, I . . . My hands are tied. The bishop, you see, has . . . He, uh . . . I'm afraid there's no nice way to say this, Professor. The bishop has banished Lutherans from this diocese.

RHETICUS. What has that got to do with me?

COPERNICUS. You mean you're not? Lutheran?

RHETICUS. I'm not looking to settle down here. I just want to talk to you, about your work.

COPERNICUS. Even that would not be . . . No.

RHETICUS. I'm a mathematician, not a theologian. Couldn't you explain that to him? Pernaps he would grant us a . . . What do you call it? An indulgence?

COPERNICUS. A dispensation. But, no. There's no chance of that.

RHETICUS. Oh, please try. You can promise him that our discussions will have nothing to do with faith. We'll limit ourselves strictly to arithmetic and geometry. The wings of the human mind. On such wings as those, we can transcend our religious differences. Transcend all religious differences. Didn't Abraham teach astronomy to the Hebrews? And Moses, another Jew? And Heaven knows, all those Islamic astronomers, praying to their Allah five times a day, then watching the stars all night. Even going back to the Egyptians, the Greeks! Prometheus and the theft of divine fire! The very crime for which he suffered an eagle to devour his liver! What does that mean, if not that Prometheus delivered the light of astronomy to mortals?

Beat.

COPERNICUS. How young you are, Professor.

RHETICUS. You're not afraid to talk to him, are you?

COPERNICUS. I? I am the bishop's personal physician.

RHETICUS. Well, then.

COPERNICUS. I was summoned to his side tonight, after he was "poisoned" by a Lutheran spy.

RHETICUS. No!

COPERNICUS. No. It was nothing like that. But knowing the intimate details of his digestion gives me no leverage to sway his opinion. On any subject.

RHETICUS. (*kneeling*) Please try! I implore you. If you do, I swear I will . . .

COPERNICUS. Come, Professor. You must leave off this genuflecting and swearing. Remember, you are not a Catholic, and I am not a priest.

RHETICUS. You're not?

COPERNICUS. Only minor orders. Never ordained. But I do administer the cathedral's business affairs. I'm an officer of the Church. I cannot harbor a heretic.

Beat.

COPERNICUS. I'm sorry if I've offended you. I meant no disrespect for your beliefs.

RHETICUS. You mean . . . I'd be a danger to you?

COPERNICUS. You are a danger to yourself, young man. Rushing off to unknown places, knocking on strangers' doors, shouting about missions and quests.

RHETICUS. I only meant to . . .

COPERNICUS. (*picking up satchel, pressing it on* RHETICUS) Now you be careful on the roads, mind you. Watch out for yourself out there.

RHETICUS. You won't let me stay after all?

COPERNICUS. I'm sorry to disappoint you.

RHETICUS. What will I do now? How will I ever . . . ? Oh, God!

COPERNICUS. If you really want to pursue my ideas, why don't you write to me? After you get back to Wittenberg, you could . . . I don't mean write directly. You would need to send your letters through an intermediary. Perhaps Schöner would agree to serve as a . . . a point of contact for us. I would like that.

COPERNICUS *goes to* RHETICUS, *puts a friendly arm around his shoulders, to shore him up.*

COPERNICUS. Now then, Professor. Gather your things. Take the books, please. I could not keep them in good conscience. I hate to send you away like this. But we are victims of these times.

RHETICUS *glumly does as he's told. Together they walk to the door. When* COPERNICUS *opens it, daylight floods the room.*

COPERNICUS. Oh, for Heaven's sake!

COPERNICUS *shuts the door and pushes* RHETICUS *back into the room.*

COPERNICUS. You can't go now!

RHETICUS. Sir?

COPERNICUS. It's too late. Look! Daylight already. I'll have to . . . Where . . . ? I know!

COPERNICUS *moves aside a bench to reveal a trapdoor, which he struggles, unsuccessfully, to open.*

RHETICUS *watches, dumbfounded.*

COPERNICUS. Help me!

RHETICUS snaps to and pulls on the trapdoor until it opens.

COPERNICUS. Now come this way. Hurry.

They disappear through the trapdoor, closing it behind them.

The stage is empty for a moment as the dawn light continues to brighten. An all-male choir can be heard chanting Matins.

Someone knocks urgently at the front door.

ANNA enters from an interior room, dressed, tying on an apron.

ANNA. Who's there?

She looks around the room, replaces the bench, tries to restore normal order, goes to the door.

FRANZ. (*entering*) Oh, Miss Anna! You're still here.

FRANZ throws his arms around her, as a child to a mother, near tears.

ANNA. What is it, pet? What's the matter?
FRANZ. You're here. You're still here.
ANNA. There, there, now. Where else would I be?
FRANZ. I don't know. Oh, Miss Anna, I don't want you to go away.
ANNA. What's got into my brave little man? Oh, poor lamb. There, there.
FRANZ. He said you had to go away.
ANNA. Who said such a thing?
FRANZ. The bishop said.
ANNA. The bishop?
FRANZ. I heard him.
ANNA. No.
FRANZ. He did. He told the doctor to make you go away. Oh, please don't go, Miss Anna! Please don't go.

Blackout.

Scene iv. The Bishop's bedroom
TWO BISHOPS

The Bishop, *in his bed, tosses in the throes of a nightmare.*

A knock at the door exaggerates the terror of his dream, but also wakes him, and he cries out.

Giese. (*offstage*) So sorry to disturb you . . .

The Bishop *slowly recognizes his surroundings, comes out of the dream.*

Giese. (*opening the door*) It's the day of our chess game, Johann. Remember? May I come in?

Bishop. (*throwing off the bedclothes, sitting up*) No.

Giese. (*entering*) What's the matter with you? Why are you still in bed?

Bishop. Did the boy let you in?

Giese. Are you ill?

Bishop. Yes. No! But I feel addled. Like a horse kicked me in the head.

Giese. We should send for Nicholas to come and examine you.

Franz *enters with a pitcher and basin, sets them on a washstand, exits.*

Bishop. Nicholas was here all night. An awful night I had. Some cursed Lutheran tried to poison me.

Giese. Poison?!

Bishop. Tried to kill me. And very nearly succeeded.

Giese. Heaven forbid.

Bishop. Agh! I ask you, Tiedemann: If I'm not safe in my own dining room, where am I safe? Lutherans everywhere. In the kitchen. In the soup.

Giese. Have you apprehended a suspect?

Bishop. I can't tell who is trustworthy anymore. I may have to torture someone to get at the truth.

The Bishop *rises, goes to the washstand, and, through the following dialogue, removes his nightshirt, grooms himself.*

GIESE. Are you sure it was poison? What did Nicholas say?

BISHOP. Nicholas! His skills may combat a single instance of poisoning. And thank God for that. But his medicaments cannot stanch the spread of the Lutheran plague. It oozes and festers all around us. As God is my witness, it has reached epidemic proportions!

GIESE. You talk like a soldier, Johann.

BISHOP. And you, Tiedemann! You sit idly by, and watch. You do nothing to stem the tide.

GIESE. What would you have me do? Lay siege to Wittenberg?

BISHOP. You have still not adopted my edict in your diocese. Have you?

GIESE. Now, Johann.

BISHOP. You won't do even that much.

GIESE. You know how I feel about . . .

BISHOP. We're the only ones left, Tiedemann. You and I. We're the last holdouts in the whole region. Every other bishop, to a man, has bowed to that devil Luther. God help us, even the duke has converted. We are surrounded. We must crush the menace.

GIESE. We are men of God, Johann.

BISHOP. The Church calls us to her defense. I need your support. As long as you allow Lutherans to live and work in Kulm . . .

GIESE. Our Lutherans in Kulm don't cause any trouble. They just . . .

BISHOP. Listen to me, Tiedemann. If we have trouble here in Varmia, you have trouble in Kulm. We have the same troubles, you and I. How do you know my assassin wasn't one of your Lutherans?

GIESE. These are peasant farmers. Merchants. Tradesmen. The same people who have lived among us for generations, since long before . . .

BISHOP. They have betrayed us, by betraying the Church. You cannot let them go about with impunity.

GIESE. In your heart, you know there's a better path to reconciliation with our Protestant brethren.

BISHOP. Oh, please, Tiedemann! When will you face the facts?!

GIESE. We're all Christians in the eyes of God.

BISHOP. Haven't the past twenty years taught you anything? That sniveling little monk! He has whined and complained and . . . and gained himself a huge following! How did it happen? Hm? Who ever thought anyone would listen to him? Now look at him. He sings a few hymns, and half the continent thinks he's the Second Coming.

The BISHOP *finishes his grooming, throws down his towel like a gauntlet.*

BISHOP. It's an abomination.

GIESE. The Church has weathered worse storms before this. If we are steadfast in our faith, and treat our fellow citizens with compassion . . .

BISHOP. You mean you refuse to back me?

GIESE. I'm saying that the changing times challenge us to summon new reserves of patience, so we can negotiate peaceably with . . .

BISHOP. You have more tolerance for Lutherans than you have for me.

GIESE. Let us pray together, for guidance. "Our Father, Who art in Heaven . . ."

BISHOP. I bet you'd just love for one of them to do away with me. So you could take my place, and be bishop here yourself.

GIESE. Don't give in to such dark thoughts, Johann. Pray with me now. "Our Father, Who art in Heaven, hallowed be Thy name . . ."

GIESE *keeps praying, whispering under the* BISHOP'S *lines, speaking louder between them.*

BISHOP. That's why you keep your canonry here, isn't it? You want to have your foot in the door, so when I die . . .

GIESE. "Give us this day our daily bread. Lead us not into temptation, but deliver us from evil. Forgive us our trespasses, as we forgive those who trespass against us . . ."

BISHOP. Why didn't I see it before? Why else would you remain a canon here in Varmia?

GIESE. "Amen."

BISHOP. You should give up your canonry!

GIESE. What?!

BISHOP. You have no right to be a canon here any longer.

GIESE. Don't be silly, Johann. I have every right . . .

BISHOP. I want you to resign. Right now. You should step down of your own volition. Don't make me force you out.

GIESE. You cannot force me to . . . It's a lifetime appointment. Everyone knows that.

BISHOP. Nevertheless, you are free to leave it.

GIESE. Why would I? I rely on my income from the canonry.

BISHOP. You're Bishop of Kulm now.

GIESE. Kulm is such a poor diocese. You know that better than anyone, Johann. When you were Bishop of Kulm . . .

BISHOP. You cannot be Bishop of Kulm and canon of Varmia, too.

GIESE. Of course I can. You did. When you were Bishop of Kulm, you didn't give up your Varmia canonry.

BISHOP. What I did has nothing to do with what you should do.

GIESE. But it's exactly the same situation. You remained a canon here the whole time you were Bishop of Kulm. If you hadn't done that, you could never have been elected Bishop of Varmia.

BISHOP. Aha! You admit it, then! You do want to take my place!

GIESE. I'm older than you, Johann. I'm not likely to outlive you.

BISHOP. Not likely, no. Except in the event of my untimely death.

GIESE. You cannot accuse me of such treachery!

BISHOP. Can't I?

GIESE. It's the principle of the thing. And the income, of course. And I . . . I still belong to this community. These are my lifelong friends. Nicholas and I go back . . .

BISHOP. Don't expect your friend Nicholas to come to your rescue now. He's on very shaky ground himself.

GIESE. Nicholas?! He keeps all of us alive!

BISHOP. I could have him excommunicated.

GIESE. Have you gone mad, Johann?

BISHOP. I refuse to look the other way any longer while that harlot comes and goes as she pleases.

GIESE. You mean the housekeeper?

BISHOP. Housekeeper, harlot. What's the difference? What do you take me for? A simpleton? You think I don't know a harlot when I see one?

GIESE. But he has trained her, Johann. About herbs and . . . Medicinal herbs, I mean. She makes . . . medicines. She . . .

BISHOP. I had no idea you were so fond of her, Tiedemann. Why don't you take her home with you? That would solve everything. The harlot goes. Nicholas is absolved. And you are content to stay home in Kulm with your Lutherans and your new concubine.

Blackout. The choir chants Lauds.

SCENE V. TOWER ROOM
WORLD MACHINE

Dim lights reveal the tower room as small and spare, dusty from disuse, almost scary, with minuscule windows and a low door. The furnishings include a table and chair, a cot, and the World Machine, a globe-like nest of intersecting rings, about the size of a manned spacecraft capsule, perched on a pedestal.

RHETICUS. (*offstage*) Where are you taking me, sir?

COPERNICUS. (*offstage*) Only a little farther now.

RHETICUS. (*offstage*) But where . . . ?

COPERNICUS. (*offstage*) We're nearly there . . . Ah!

COPERNICUS *enters, breathless, with a lantern.*

COPERNICUS. Here we are.

RHETICUS *follows him in, stays close.*

COPERNICUS. You can stay here.

RHETICUS. Here?

COPERNICUS. This is it.

RHETICUS. What is this place?

COPERNICUS. You'll be safe here.

RHETICUS. Is it your observatory?

COPERNICUS. This? No.

RHETICUS. Not a prison cell, is it?

COPERNICUS. Oh, no. It's a retreat. A safe house. We all have rooms like this. When there's danger, from outside, we come up here, and . . . and we stay here until . . . until it's safe to leave.

RHETICUS. You really expect me to stay here?

COPERNICUS. No one will think to look here now. In peacetime.

RHETICUS. For how long?

COPERNICUS. Just till tonight.

RHETICUS. The whole day?!

COPERNICUS. After sunset, you can go. As soon as it's dark, I'll come fetch you.

RHETICUS. You're not staying with me?

COPERNICUS. No.

RHETICUS. But now we have all day.

COPERNICUS. I can't stay with you. I have to . . .

RHETICUS. Oh, please. Stay and seize this day with me. Look how God has provided a space of time for us, after all. This is our chance to talk. One mathematician to another. I . . .

RHETICUS *sees the Machine.*

RHETICUS. What's that?

COPERNICUS. That?

RHETICUS. What is it?

COPERNICUS. Just . . . something I made.

RHETICUS. You built it?

COPERNICUS. A long time ago.

RHETICUS. But what is it? Some kind of observing instrument?

COPERNICUS. No. No, it's . . . more of a model, really.

RHETICUS. Like an armillary sphere?

COPERNICUS. You might say.

RHETICUS. Only larger.

COPERNICUS. Yes.

RHETICUS. Much larger.

COPERNICUS. I don't use it anymore.

RHETICUS. Why so big?

COPERNICUS. Well, the person inside needs room to . . .

RHETICUS. There's someone inside it?!

COPERNICUS. Not now.

RHETICUS. No. But a person could . . . ?

COPERNICUS. Yes. The person has to sit inside it, to get the effect.

RHETICUS. And what effect would that be? Inside?

COPERNICUS. The sense of . . . the consequences, really, of my theory.

RHETICUS. So, you sat in there, while you were figuring out how to . . . ?

COPERNICUS. No. I stood out here, to operate it.

RHETICUS. Someone else was inside?

COPERNICUS. Yes.

RHETICUS. So you did have a student? Before me?

COPERNICUS. No.

RHETICUS. Then why . . . ?

COPERNICUS. No, I made this for my . . . for a friend. Someone who couldn't grasp the mathematical concepts. Who needed a way to . . . visualize the spheres.

RHETICUS. You certainly went to a lot of trouble.

COPERNICUS. I suppose I did.

RHETICUS. For your friend.

COPERNICUS. Yes. Well, then. You wait here, and . . .

RHETICUS. Could I try it?

COPERNICUS. No, I don't think so.

RHETICUS. I'd really like to see what it does.

COPERNICUS. No one's used it in years. I doubt it still works.

RHETICUS. Let's try it and see.

COPERNICUS. There's no need. You, of all people, can follow the math.

RHETICUS *fumbles about the Machine, looking for a way in.*

RHETICUS. I was hoping to read your work, sir. I didn't know I could ride in it.

COPERNICUS. Don't touch that.

RHETICUS. How do you get in?

COPERNICUS. Not there. No, not like that.

RHETICUS. Show me, then. Please.

COPERNICUS. Let go of that. It's over here. You climb in through here.

RHETICUS *dives in, but finds entry a struggle.*

RHETICUS. This is a lot smaller than it looks. There's hardly room to . . . Your friend must have been half my size.

COPERNICUS. Maybe you shouldn't . . .

RHETICUS. All right. I'm in.

COPERNICUS *closes the hatch.*

The lights suddenly go out.

RHETICUS. Oh, my God! What happened?

COPERNICUS. It will take a moment to . . .

Very dim lights come up, just enough to show RHETICUS *inside the Machine.*

RHETICUS. It's pitch-black in here. I can't see my hand in front of my face.

COPERNICUS. I'm lighting it now . . . Just another moment . . .

RHETICUS. Stuffy, too. I can hardly breathe.

COPERNICUS. There!

Little twinkling star lights appear, as in a planetarium.

RHETICUS. What the . . . ?

COPERNICUS. Do you see anything?

RHETICUS. Oh, my God!

COPERNICUS. You see?

RHETICUS. Oh, dear God. It's . . . There are stars everywhere. All around. How did you do that?

COPERNICUS. Now I turn you.

RHETICUS. What?

COPERNICUS. I said, I'll turn you around now.

RHETICUS. I can't hear you.

COPERNICUS. (*grunting with effort, muttering*) Turn. I have to turn . . . ugh . . .

RHETICUS. Oh, turn. Is that what you said?

COPERNICUS. Ugh . . . It's stuck. Wait . . .

RHETICUS. Should I do something?

COPERNICUS. No . . . Ugh . . . Ah, there it goes!

RHETICUS *is rotated in his seat. He continues rotating slowly through the following dialogue.*

RHETICUS. Good God! What's happening? Oh, this is . . . This is unbelievable.

COPERNICUS. You see? What it does?

RHETICUS. Oh, sir! You have reproduced the night. The effect is . . . It's so lovely. So . . . Oh, look! The zodiac constellations.

COPERNICUS. The thing to look for . . .

RHETICUS. There's the Ram, the Bull . . .

RHETICUS'S *seat gives a lurch.*

RHETICUS. Whoa! What was that?

His seat speeds up.

RHETICUS. Oh!

COPERNICUS. I'm sorry.

RHETICUS. Merciful heavens!

COPERNICUS. Something slipped. I'm trying to . . .

RHETICUS'S *seat slows down.*

RHETICUS. Whew!

COPERNICUS. I'm sorry. I told you the machine was . . .

RHETICUS. Oh, please, continue.

COPERNICUS. I'm afraid I can't keep this up much longer.

RHETICUS'S *seat slows down more.*

COPERNICUS. Not as strong as I used to be.

RHETICUS'S *seat slows to a stop. The stage lights return.*

COPERNICUS. You'd better come out now.

RHETICUS. (*emerging from the Machine, wobbly*) Oh, my. That was . . .

COPERNICUS. Steady, there.

RHETICUS. I'm still seeing stars.

COPERNICUS. Let your eyes adjust to the light.

RHETICUS. Ooh.

COPERNICUS. Well?

RHETICUS. Thank you, sir.

COPERNICUS. Did you find it convincing?

RHETICUS. Convincing?

COPERNICUS. Did you?

RHETICUS. Convincing of what, sir?

COPERNICUS. Of the motion.

RHETICUS. Oh, most definitely.

COPERNICUS. Good. Well, then.

RHETICUS. All the stars moved. I could see them spinning round and round.

COPERNICUS. No, the stars didn't . . .

RHETICUS. It was great.

COPERNICUS. That was you going around. Not the stars.

RHETICUS. No, I saw the . . . The stars turned around me.

COPERNICUS. You turned. In that little seat. That's the only part that moves.

RHETICUS. But I didn't feel it move.

COPERNICUS. You're not supposed to.

RHETICUS. No?

COPERNICUS. No. That's just it. You think the stars are turning, but really it's you turning. Well, it's really me turning you. And once you realize that it's you going around, then you make that shift in perception. You see?

RHETICUS. I'm not sure I do. No.

COPERNICUS. The machine gives you a physical appreciation. For what the turning of the Earth . . . You know: how the Earth, by its rotation, makes the stars appear to spin around it. And the planets, too. I tried to build in the planetary effects . . . the stations and retrogrades . . . but I had trouble aligning them.

RHETICUS. Do you mean to say . . . ?

COPERNICUS. I think those parts must still be around here, somewhere . . .

RHETICUS. Oh, no!

COPERNICUS. What?

RHETICUS. You mean, you really do mean to turn the Earth?

COPERNICUS. You knew that.

RHETICUS. But . . . really turn it?

COPERNICUS. What did you think?

RHETICUS. I didn't think you meant to turn it . . . physically.

COPERNICUS. How else would it turn, if not physically?

RHETICUS. It would turn . . . theoretically. You know. In a hypothetical way. On paper. In order to . . .

COPERNICUS. No.

RHETICUS. Theoretically. Mathematically. But not . . .

COPERNICUS. No, the motion is real. Of course it is.

RHETICUS. Oh, my God.

COPERNICUS. I thought you understood my work.

RHETICUS. I . . .

COPERNICUS. Didn't Schöner explain it to you?

RHETICUS. He, uh . . . I . . .

COPERNICUS. What did he tell you?

RHETICUS. I don't think he sees it quite the way you do, sir.

COPERNICUS. How can that . . . ?

RHETICUS. He didn't mention anything about a real motion.

COPERNICUS. Are you sure?

RHETICUS. All he said was . . . No, he didn't say anything about . . .

COPERNICUS. You mean, he doesn't understand it either?!

RHETICUS. I think he must not have interpreted it . . . literally.

Beat.

RHETICUS. Why would he?

COPERNICUS. Why?!

RHETICUS. Why would he leap to that conclusion?

COPERNICUS. Oh, dear God!

RHETICUS. Honestly, sir, I don't think anyone realizes exactly what it is that you have in mind.

COPERNICUS. What can they think I've been doing all these years?

RHETICUS. Even just to . . . to use the idea as the basis for new calculations, would . . . But, to claim the motion as reality?!

COPERNICUS. Yes.

RHETICUS. I am . . . dumbstruck. I . . . Look! You and I. We're just standing here. The Earth . . .

RHETICUS *stamps his foot a few times to make the point.*

RHETICUS. It doesn't move.

Beat.

COPERNICUS. Yes it does.

RHETICUS. You really believe the Earth is . . . turning?

COPERNICUS. It's not a question of belief, Professor. I know it turns.

A peal of bells begins, continues through the following dialogue.

RHETICUS. What do you mean, you "know"?

COPERNICUS. I mean the evidence has convinced me.

RHETICUS. What evidence?

COPERNICUS. (*hearing the bells*) Goodness, the time!

RHETICUS. You mean the Earth leaves some kind of wake behind it? Like a boat?

COPERNICUS. I'm sorry. I must leave you now.

RHETICUS. No, wait a minute.

COPERNICUS. You must excuse me. I'll come back tonight.

RHETICUS. Wait!

COPERNICUS. They're expecting me in the . . .

RHETICUS. Just because I raise a few questions? You walk away?

COPERNICUS. Don't you hear the bells? That's the call to Mass. If I'm not seen in the cathedral, then . . .

COPERNICUS *exits.*

RHETICUS. Wait!

The bells continue, grow louder.

RHETICUS *stares at the door, incredulous at what he's just heard, also furious and afraid. He glares at the Machine, grabs hold and shakes it.*

Blackout. Pealing turns into tolling of the hour: 12 o'clock.

SCENE vi. INSIDE COPERNICUS'S HOUSE
ILLNESS

About two hours later, GIESE *lets himself into the house.*

GIESE. Hello? Nicholas? Are you here?

ANNA. (*offstage*) Mikoj?! Oh, thank goodness. I've been so worried all this . . . (*entering*) Oh! Oh, forgive me, Reverend Father.

GIESE. He's not here?

ANNA. No, Your Reverence.

GIESE. I was supposed to meet him after Mass.

ANNA. He wasn't at Mass?

GIESE. Of course he was at Mass.

ANNA. Yes, of course. Of course he was.

GIESE. And now?

ANNA *bows her head to hide her face.*

GIESE. I understand. What a difficult time this must be for you.

ANNA. Your Reverence, I . . . May I confide in you?

GIESE. You wish to make a confession?

ANNA. No, just . . . just to let you know something. A secret. As a good and loyal friend of this house.

GIESE. You need not tell me anything.

ANNA. Last night, I . . .

GIESE. Now, now, you mustn't take all the blame on yourself. It's never one-sided in these situations. I know that. To be frank, I feel I am partly responsible. I've known about it all along. And yet I said nothing. As Nicholas's friend, I should have counseled him. I could have saved him from this . . . this ridiculous threat. But don't fret. Nothing bad will happen to him if you are brave and do what's required of you. Tell me, do you have family who could take you in?

Beat.

GIESE. Or a friend, perhaps? Someplace where you know people, where you'll feel welcome?

A scuffling sound comes from under the floor, at the trapdoor, and something bumps against it from below.

ANNA *jumps, cries out in fright.*

GIESE. What was that?

Another thump sounds from the trapdoor.

COPERNICUS. (*offstage; panting, whispering*) Anna?

GIESE. Good heavens!

COPERNICUS *is trying to lift the trapdoor from below.*

GIESE. There's someone in the passage.

COPERNICUS. (*offstage*) Tiedemann? Is that you? Let me up.

GIESE *moves the bench, lifts the door with great difficulty, as* COPERNICUS *pushes it from below.*

COPERNICUS, *panting, drags up the unconscious body of* RHETICUS.

ANNA *screams, then recovers herself and moves to help. The three of them pull* RHETICUS *into the room.* ANNA *puts her shawl under his head, touches his face.*

ANNA. He's burning up with fever.

COPERNICUS, *exhausted from the effort, sits on the floor near* RHETICUS.

RHETICUS *shakes with chills, moans.*

GIESE. Who is this?

ANNA. I'll get some blankets.

COPERNICUS. And willow bark.

ANNA. (*exiting*) I know.

GIESE. Poor fellow. What's wrong with him?

COPERNICUS *continues to catch his breath.*

GIESE. Oh, never mind. You can tell me later. But what were you doing there, Nicholas?

COPERNICUS. (*squeezing* GIESE's *hand*) So good to . . .

GIESE. I know. I won't ask you any more questions now . . . Goodness, I haven't been up there since . . .

Beat.

GIESE. Do you still have your . . . your machine, with all the . . . ?

ANNA. (*returning with blankets, water*) Why didn't you leave him in the tower?

COPERNICUS. Too sick.

COPERNICUS *holds up* RHETICUS's *head, pours a few drops of medicine into his mouth.*

GIESE. What's wrong with him?

COPERNICUS. Ague. Exposure. God only knows where he's slept in weeks of travel.

GIESE. Where did he come from?

ANNA. His clothes are drenched with sweat.

COPERNICUS. Better get them off.

ANNA *and* COPERNICUS *undress* RHETICUS, *wrap him in blankets, through the following dialogue.*

RHETICUS *shakes with chills, moans, resists them mindlessly.*

ANNA. How will you explain this . . . ?

COPERNICUS. I'll think of something.

ANNA. You should never have let him in the house.

GIESE. What's going on here?

COPERNICUS. I'll say I was coming home, from the bishop's, late at night, when I found him, lying in front of my house.

ANNA. You can't say you . . .

COPERNICUS. That much is true. He was ill. How could I leave him out there, weak and sick?

GIESE. You dragged him into the house? And up to the tower? And then back down from the . . . ?

COPERNICUS. No, Tiedemann. He walked into the house. And then . . . We had to . . . But I did find him lying out there. So I took him in.

ANNA. A total stranger?

COPERNICUS. It was the Christian thing to do.

ANNA *shakes her head, continues tending to* RHETICUS.

GIESE. I would have done the same. But he's much better off here, with you. It was his great good fortune that Providence delivered him to your door, Nicholas.

COPERNICUS. That's it! Providence delivered him. So that I could care for him in his hour of need.

ANNA. But, a Lutheran?

GIESE. What?

COPERNICUS. How did I know? He was unconscious.

GIESE. He's a Lutheran?

COPERNICUS. Later he became delirious. It was impossible to make any sense of what he said. We still have no idea who he is. Or where he came from. All his papers had been stolen. By robbers. Highwaymen.

ANNA. Why are you protecting him?

COPERNICUS. Anna, please. Make up a bed for him in the pantry. There's nothing else to be done until his fever comes down.

ANNA, *still disapproving, goes to the pantry as told.*

COPERNICUS *slumps, head in hands.*

GIESE *goes to him, pats and rubs his shoulders.*

GIESE. All right, my friend. From the beginning now. Who is this prodigy among us?

COPERNICUS. Of all the times for someone like him to . . . Someone of his talents . . . Why now? Agh! If only he'd come to me twenty years ago.

GIESE. Twenty years ago he was still in swaddling clothes, from the look of him.

COPERNICUS. It wouldn't have made a difference then either. My ideas are too disturbing to see the light of day.

Beat.

GIESE. He came to you about that?

COPERNICUS. So he said.

GIESE. What about it?

COPERNICUS. Nothing. It doesn't matter. He didn't really understand it anyway.

GIESE. But he traveled here? To find you?

COPERNICUS. Incredible, isn't it?

GIESE. From where?

COPERNICUS. You wouldn't believe me if I told you.

GIESE. What did he say?

COPERNICUS. Came all this way. With letters from . . . He had a letter of introduction from Schöner.

GIESE. Nuremberg Schöner?

COPERNICUS. And Hartmann, too. And a stack of books he wanted to give me. Ptolemy in the original Greek. Can you imagine? And here. Look at these.

COPERNICUS *gives* GIESE *the notes.*

GIESE. What's this?

COPERNICUS. You and I never saw Mercury at an angle of western elongation like that. Nowhere even close to those values.

ANNA. (*returning*) It's ready.

All three pick up RHETICUS *and carry/drag him toward the pantry.*

FRANZ *enters, unnoticed by the others, takes in the scene.*

COPERNICUS *and* GIESE *return.*

COPERNICUS. How long have you been here, lad?

FRANZ. I . . . His Reverence sent me, Doctor.

COPERNICUS. Did you see . . . ?

FRANZ. His Reverence wishes Bishop Giese to attend him in his chambers, to witness the signing of the edict.

COPERNICUS. (*to* GIESE) He's gone and done it? Already?

GIESE *takes a last, appreciative look at the observations.*

GIESE. One thing is certain, Nicholas. The Lord surely works in mysterious ways.

GIESE *gives back the notes to* COPERNICUS, *exits with* FRANZ.

ANNA *returns.*

COPERNICUS. The bishop's boy was here.

ANNA. Again?

COPERNICUS. Do you think he saw anything?

ANNA. What did he overhear, Mikoj? Between you and the bishop?

COPERNICUS. He told you about that?

ANNA. It's true, then? (*rushing into his arms*) Oh, Mikoj!

COPERNICUS. (*embracing her*) He was overwrought last night. Sick and fearful. He'll forget about us.

ANNA. Bishop Giese said something to me about . . .

COPERNICUS. No, no. Hush.

ANNA. Yes, he did. He asked me where I was going. And did I have family to take me in.

COPERNICUS. Don't worry, dearest.

ANNA. Oh, Mikoj!

COPERNICUS. I won't let anything happen to you.

ANNA. He can't really make you send me away? Can he?

COPERNICUS. He'll have to kill me first.

They kiss, continue to hold each other.

ANNA. I won't go. I won't leave you, Mikoj.

COPERNICUS. I won't let you.

RHETICUS *cries out from the other room.*

COPERNICUS *and* ANNA *turn, start in his direction, but he quiets, so they stay where they are, clinging to each other.*

Blackout.

SCENE vii. BISHOP'S PARLOR
PLEA BARGAIN

The BISHOP *sits at the desk where he signs and seals the edict.* FRANZ *stands behind him,* GIESE *facing him.*

BISHOP. He just took him in? Without even knowing his identity?

GIESE. That's Nicholas for you. If he sees a person is sick, he simply acts.

BISHOP. But this fellow could be a spy, for all he knows.

GIESE. No, he's a mathematician.

BISHOP. I thought you said no one knew anything about him.

GIESE. That's right. No papers. But he had several books. In his travel bag.

BISHOP. Books in a bag don't prove a person's profession.

GIESE. These were large textbooks, about mathematics. That the robbers did not take.

BISHOP. No wonder.

GIESE. I think he came here on purpose, Johann. Expressly to engage Nicholas about his theory. To shake him out of his paralysis.

BISHOP. So what if he did? What of it?

GIESE. Think what it would mean, Johann. You know how I've always said one day Nicholas will bring glory to Varmia through his mathematical work.

BISHOP. That is one harebrained idea, that theory of his. I thought he was wise to put it aside.

GIESE. He should be encouraged to take it up again.

BISHOP. He should let it lie. It's a dangerous notion.

GIESE. It's controversial, I grant you, but . . .

BISHOP. It may even be heretical.

GIESE. Oh, no, Johann.

BISHOP. Then it's a laughingstock. You should hear what they used to say about him at court. How he mistook the Earth for a side of beef. So he put it on a spit, and tried to roast it in the Sun's fire.

GIESE. His ideas are beyond the comprehension of ordinary minds like yours and mine.

BISHOP. Even mathematicians have common sense, Tiedemann. Now, then. Stop changing the subject. And add your name to this document. Will you do that? Will you stand with me to protect Varmia? And Kulm. And the rest of our province, and Poland, and the world, from a clear and present danger?!

GIESE. I cannot condone the punishment of innocent people.

Beat.

BISHOP. I have already written my recommendation to the provost of the chapter, requesting that you be relieved of your canonry. I have it right here, just waiting for my signature and seal. You sign the edict, Tiedemann, and I'll tear up the letter.

GIESE. I must be getting back to Kulm now.

BISHOP. Sign, damn it!

GIESE. I have preparations to make, to receive my guests. I've invited Nicholas to bring his unfortunate visitor to Kulm, as soon as the youth is well enough to travel.

BISHOP. The sooner he leaves here, the better.

GIESE. And the nurse, to look after him until he's completely . . .

BISHOP. Good riddance.

GIESE. And Nicholas, of course.

BISHOP. Nicholas isn't going anywhere.

GIESE. He will leap at the chance to engage another mathematician in learned discourse.

BISHOP. You can have the stranger. But I won't let you take Nicholas that far away.

GIESE. How I shall enjoy hearing them discuss the wanderings of the planets through the visible heavens, while I tend to the invisible one.

BISHOP. I need him here with me. He belongs to me.

Blackout.

SCENE viii. COPERNICUS'S HOUSE
ASTROLOGY

COPERNICUS *and* ANNA *huddle together in an embrace, as before; they jump when . . .*

RHETICUS *staggers in, wrapped in a blanket.*

ANNA. Good God!

RHETICUS. What happened? Why didn't you tell me?

COPERNICUS. What . . . ?

RHETICUS. It's dark now. Can't you see? It's dark!

RHETICUS *stumbles, starts to fall.*

ANNA *and* COPERNICUS *catch him, sit him down.*

COPERNICUS. Bring him some of that broth.

ANNA *exits.*

RHETICUS. You promised you'd tell me when it got dark.

COPERNICUS. You're ill. Do you remember? You're not going anywhere tonight.

RHETICUS. Where are my clothes?

COPERNICUS. (*taking off his cassock, putting it around* RHETICUS) You're still weak. You need to . . .

RHETICUS. I can't stay here.

RHETICUS *tries to stand up, falls back into the chair.*

COPERNICUS. In another day or two, you'll be stronger. Then you can do as you please. But for now you're in my care.
RHETICUS. This is your house. We were in this room.

ANNA *enters, with a cup.*

COPERNICUS. Here, drink this.
RHETICUS. But this isn't where we . . . We went somewhere else to . . .
COPERNICUS. Go on, drink it. It's good for you.

ANNA *goes to the room where* RHETICUS *was resting.*

RHETICUS. You put me in that . . . machine.
COPERNICUS. Drink this, now. It's full of medicine.
RHETICUS. (*taking the cup, then dropping it*) Oh, *no*!
COPERNICUS. It's all right. There's more where that came from.
ANNA. (*returning, with* RHETICUS'S *clothing*) His clothes are still wet.
COPERNICUS. Please brew some more broth for him.
ANNA. (*exiting*) I'll hang these by the kitchen fire.
RHETICUS. Now I remember. Oh, no. Oh, no, no, no.
COPERNICUS. You must have been dreaming.
RHETICUS. I thought you would save me.
COPERNICUS. Sometimes fever causes very vivid, frightening dreams.
RHETICUS. You! I thought you could help me.
COPERNICUS. I've done everything I know how to . . .
RHETICUS. What will I do now?
COPERNICUS. You'll be fine.
RHETICUS. What will become of me?
COPERNICUS. Good as new, you'll see.
RHETICUS. I came here in good faith . . .
COPERNICUS. Yes, yes. I know.
RHETICUS. And what do I find? A lunatic! A deluded old . . . a . . . a recluse! Obsessed with an insane idea.
COPERNICUS. Get hold of yourself, now.

RHETICUS. (*jumping up, stronger now*) Where are my clothes? Where's my satchel?

COPERNICUS. You don't need any of that now.

RHETICUS. My horoscopes are in there.

COPERNICUS. I don't have to see your horoscope. I know how to treat your symptoms without that.

RHETICUS. You don't understand. Where is that satchel?

COPERNICUS. Calm yourself.

ANNA. (*returning with another cup*) You must have left his things up in the tower.

COPERNICUS. They can wait there for now. Here.

RHETICUS *resists, but then weakens again, drinks the broth.*

ANNA. I think we all need something to eat.

ANNA *exits.*

RHETICUS. I know it by heart. I can recite the whole thing without looking at it.

COPERNICUS. What can you recite?

RHETICUS. Every house, every aspect, every conjunction and opposition. Every indicator of doom.

COPERNICUS. Don't tell me you believe in that?

RHETICUS. It's not as though I have a choice.

COPERNICUS. You should know better.

RHETICUS. If only I could forget what I know.

COPERNICUS. (*a little sarcastic*) Change it, then. If you don't like what your horoscope portends, you can simply reconfigure it. Isn't that right? Reapportion the houses, or adjust the presumed time of birth, and . . . make it say something else. Something better. Whatever you like.

RHETICUS. (*dead serious*) I've tried that. Tried all those things. It always comes out the same.

COPERNICUS. I'm sorry, Professor. I can't help you with your horoscope.

RHETICUS. And you call yourself a mathematician?

COPERNICUS. What do you take me for? A fortune-teller?

RHETICUS. The fates of empires depend on the positions of the planets.

COPERNICUS. No, Professor. The fates of empires depend on the positions of armies on battlefields. Not the planets in the heavens. The sky does not enter into human affairs.

RHETICUS. You don't understand.

COPERNICUS. A man's fate is in God's hands.

RHETICUS. Tell that to your pope! Don't you know he brought his favorite astrologer to Rome?!

COPERNICUS. Doesn't your Luther denounce the whole practice?

RHETICUS. I told you, he knows nothing about mathematics.

COPERNICUS. Is that all you came here for? Some new trick for casting your horoscope?

RHETICUS. Not just mine! Yours. Schöner's. Everybody's! Wars. Floods. Plagues. All the global predictions for the coming year. For years to come! That's what I saw as the fruit of your labors. The long march of history. The rise of Luther. The fall of Islam. The Second Coming of Our Lord Jesus Christ!

COPERNICUS. I give you the true order of the planets. The workings of the whole heavenly machinery, with every one of its former kinks hammered out. But all of that is useless to you, unless it provides excuses for every petty human failing.

RHETICUS. You think you can just twirl the Earth through the heavens like some . . . like a . . . like . . . Oh, my God. Wait a minute. If the Earth moved . . . then . . . If the Earth moved through the heavens . . .

COPERNICUS. It does move.

RHETICUS. If the Earth moved among the planets, then it would approach them and recede from them, and maybe even . . . It would! Yes! If that happened, it would magnify the effect of every planetary influence.

COPERNICUS. No.

RHETICUS. That would have to happen, as a natural consequence. An enhancement of the influence that each planet exerted on the individual . . .

COPERNICUS. The one thing has nothing to do with the other.

RHETICUS. How can you be sure? Have you checked for those effects?

COPERNICUS. No.

RHETICUS. Not even in your own chart? That would be so easy to do. To compare, say, Mars at opposition with Mars at solar conjunction, and then to . . .

COPERNICUS. No!

RHETICUS. This is better than I'd hoped. Better than I ever dreamed! Think what it means! This truly could dispel the whole fog of absurdity that hangs over your theory.

COPERNICUS. If you want to know the future, you should go slaughter a goat and examine its entrails. And leave the planets out of your predictions.

RHETICUS. I think there's really something to it. Let's say, just for the moment, just for argument's sake, that the Earth . . . turns. How fast would it . . . ? It has to spin around very fast, right? For the turning to cause day and night?

COPERNICUS. It is rapid, yes.

RHETICUS. How rapid?

COPERNICUS. You do the math.

RHETICUS. All right. The circumference of the Earth is . . . What? Twenty thousand miles?

COPERNICUS. Twenty-four.

RHETICUS. Twenty-four thousand, right. And it has to make a full rotation every . . . twenty-four hours.

COPERNICUS. Not a very difficult calculation, is it?

RHETICUS. God in Heaven! A thousand miles an hour?

COPERNICUS. That's what it must be.

RHETICUS. But that can't be. We would feel that.

COPERNICUS. No. We don't feel it.

RHETICUS. We don't feel it because we don't really turn.

COPERNICUS. We don't feel it because we move along with it. Like riding a horse.

RHETICUS. When I ride a horse, I feel it.

COPERNICUS. On a ship, then. Sailing on a calm sea. You move along in the direction of the wind, but you don't have any sense that you're moving.

RHETICUS. Yes, I do. I see the shore receding. I feel the breeze in my face.

COPERNICUS. Go inside the cabin, then.

RHETICUS. (*crestfallen again*) It won't work. It's too . . . It's . . . If the Earth turned as fast as you claim, there would be a gale, like the wind from God, howling and blowing against us all the time.

COPERNICUS. No, there's no wind.

RHETICUS. That's what I'm saying.

COPERNICUS. There's no wind because the air turns along with the Earth.

RHETICUS. The air? Turns?

COPERNICUS. It's all of a piece, yes. They turn together, as one. The Earth and the air. And the water, of course.

RHETICUS. We could not be moving that fast and not feel anything. It's impossible.

COPERNICUS. (*grabbing* RHETICUS *by the shoulders to shake him*) It's turning! All the time, it's turning. And that turning is what makes the Sun appear to rise . . .

COPERNICUS *turns* RHETICUS *by the shoulders, roughly, so he faces away (his back to* COPERNICUS*).*

COPERNICUS. And set . . .

COPERNICUS *turns* RHETICUS *the rest of the way around, so they face each other again.*

COPERNICUS. And rise again on the following day.

COPERNICUS *holds* RHETICUS *there for a moment, their faces close, then pushes him away, drops his hands, steps back.*

RHETICUS. What about the other motion?

COPERNICUS. You think I don't know it sounds crazy? Do you have any idea how long it took me to accept it myself? To go against the judgments of centuries, to claim something so . . . so totally at odds with common experience?

RHETICUS. Tell me about the other motion, around the Sun.

COPERNICUS. It's the same thing. You don't feel it. It's part of you, like breathing.

RHETICUS. No, I mean, is it . . . just as fast?

COPERNICUS. Oh.

RHETICUS. Is it?

COPERNICUS. No.

RHETICUS. Good.

COPERNICUS. It's faster.

RHETICUS. Damn!

They turn away from each other.

RHETICUS. (*turning back to* COPERNICUS) How fast does it go?

COPERNICUS. Around the Sun?

RHETICUS. Around the Sun, yes.

COPERNICUS. I don't know.

RHETICUS. Oh, come on. Tell me.

COPERNICUS. (*turning to* RHETICUS) I really don't know. No one knows the actual distance that the Earth would have to go to get all the way around, but it must be in the millions . . . It must be many millions of miles. Which means we go around the Sun at least . . . at *least* ten times faster than we spin.

RHETICUS. So, ten thousand miles per . . .

COPERNICUS. Maybe a hundred times faster.

RHETICUS. A hundred times a thousand miles?

COPERNICUS. Maybe.

RHETICUS. That's where it all falls apart.

COPERNICUS. Why?

RHETICUS. *Why?*

COPERNICUS. Why does it make more sense for the Sun to go around the Earth? The Sun should stand as a light for all creation, unmoved, at the center of the universe. The way a king or an emperor rules from his throne. He doesn't hurry himself about, from city to city. Once you let the Sun take his rightful place at the hearth, the Earth and the other planets arrange themselves in perfect order around it. And they take their speed from his command. That is why Mercury, the nearest to him, travels around him the fastest. And after Mercury, each successive planet takes a slower course, all the way out to Saturn, the slowest of them all.

RHETICUS. Really? They line up like that? In order of their speed?

COPERNICUS. It's as though they draw some kind of motive force from the Sun's light.

RHETICUS. What could it be? What kind of force?

COPERNICUS. I don't know. I am still in the dark on that matter. But it's there. And that's why all their motions are interrelated, as though linked together by a golden chain. You could not alter a single one, even so much as a fraction of an inch, without upsetting all the rest.

RHETICUS. The way you talk. It's as though you know God's plan.

COPERNICUS. Why else would you study mathematics? If not to discover *that*?

Beat.

RHETICUS. And the stars?

COPERNICUS. The sphere of the stars, like the Sun, also holds still. It cannot spin around the Earth every day. It's too big.

RHETICUS. I'm trying to see it your way. Really, I am. But if the Earth moves around the Sun . . . Shouldn't we see some change in the stars? Wouldn't some of them look . . . I don't know . . . closer together sometimes, or farther apart? There should be a change, from spring to fall, that people who paid attention would notice.

COPERNICUS. You would think that would happen.

RHETICUS. I don't know what to think.

COPERNICUS. But no. You don't see any seasonal difference. Because the stars are so much farther away than anyone has imagined. The scale of the universe is all but inconceivable. The distance to the stars is so tremendous that it dwarfs the distance between the Earth and the Sun. Compared to the distance from Saturn to the stars? The distance from the Earth to the Sun is . . . negligible.

RHETICUS. Negligible?

COPERNICUS. It shrinks to just a point, really.

RHETICUS. You're making this up. It's your own fantasy. The stars get in your way? You just wave them off to some other place.

COPERNICUS. Don't impose any puny, human limits on Creation. As though the whole cosmos were just a crystal ball for your own little personal affairs.

RHETICUS. In the name of the Creator, then: What is the use of all that empty space between Saturn and the stars?

COPERNICUS. The *use*?

RHETICUS. Yes.

COPERNICUS. What is the use of grandeur? Of splendor? Of glory? Thus vast, I tell you, is the divine handiwork of the one Almighty God!

Brief blackout in which the planetarium effect returns, spins, then disappears. End of Act I.

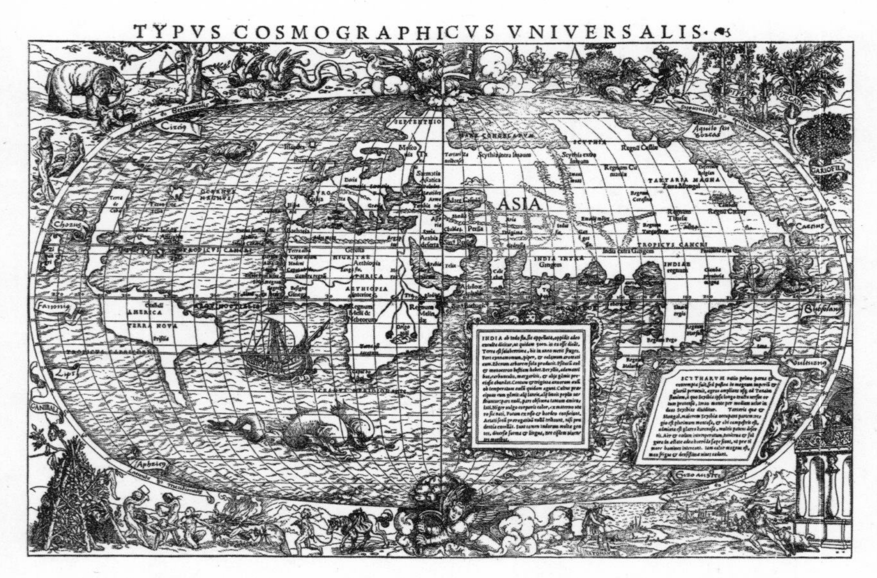

Word of Copernicus's astonishing theory, first released around 1510, challenged scholars to consider a world in motion. Almost a decade before Rheticus arrived in Frauenburg, this map became the first to suggest the Earth's rotation around a central axis, powered by cherubs turning crank handles at the poles. Published in Basel in 1532, the map has been variously attributed to Sebastian Münster and Hans Holbein.

ACT II

Scene ix. Tower room
COLLABORATORS

Very little time has passed since the preceding scene. COPERNICUS *riffles through a tall stack of pages, looking for a certain section.*

RHETICUS *still wears* COPERNICUS'S *cassock, which is too big for him.*

RHETICUS *grabs at random pages and reads them with growing excitement.*

RHETICUS. I can't believe you did all this work yourself.

COPERNICUS. I want you to see the section on Mercury. I've always known there was something wrong with my value for the anomaly. Maybe now, with Schöner's observations to add to . . .

RHETICUS. How long did it take you to make this many observations?

COPERNICUS. Here it is.

RHETICUS. There's enough work here to fill a lifetime. Have you really had no help at all?

COPERNICUS. Look at the size of the second epicyclet. The first one is on the deferent.

RHETICUS. You used *two*?!

COPERNICUS. I had to introduce the second one. Without it, the diameter of the deferent . . .

RHETICUS. Oh, now I see what you . . . Oh, yes.

COPERNICUS. The other way, with just a single . . . Here. I assembled all the correction factors in these tables. It's fairly obvious how to use them . . . But in the case of Mercury . . .

RHETICUS. I want a copy of these tables. I must have them.

COPERNICUS. Even with the tables, you still need to add in the . . .

RHETICUS. No one has a resource like this. What you've done here is . . . It's nothing short of extraordinary. It's more than the intellect or the labors of a single individual could accomplish. And yet you have accomplished it.

RHETICUS *keeps examining the manuscript, exclaiming.*

COPERNICUS *stands back, watching him.*

RHETICUS. Why haven't you published this?

COPERNICUS. You know why.

RHETICUS. Not the theory. But just these sections. Mathematicians would give anything for . . .

COPERNICUS. I don't want to divide the work that way. And pretend I don't know what I know.

RHETICUS. All right, then. The whole thing. Publish it all. Why not?

COPERNICUS. Now you sound like the crazy one.

RHETICUS. No, this is . . . This should be published. It will cause a sensation.

COPERNICUS. I will be laughed off the stage.

RHETICUS. It's all in the way you present it. Certainly you could not start off by insisting that the Earth moves.

COPERNICUS. But I would. I would have to say that.

RHETICUS. No, that will just raise everyone's hackles. You can say it later. First you show them all these other . . . Where is that . . . ? The first thing you showed me. Here! Here's a perfect example. This part, where you explain how you approached the equant problem. My God! People have been trying to solve that for . . .

COPERNICUS. No, really. It's not for publication.

RHETICUS. You must publish it.

COPERNICUS. I think Pythagoras had the right idea, when he kept his secret numbers a secret. He never divulged them to anyone, except his kinsmen and friends. And even then, only by word of mouth. Never in writing.

RHETICUS. He was afraid somebody would steal his idea.

COPERNICUS. No.

RHETICUS. You will have your name on the title page of your book.

COPERNICUS. That's not what he was afraid of. Believe me, I know how he felt. He wanted to protect his most beautiful ideas from ridicule.

Beat.

RHETICUS. You know those books I brought you?

COPERNICUS. You want them back?

RHETICUS. I met that printer, in Nuremberg.

COPERNICUS. I said you could have them.

RHETICUS. No, listen, please. He's a friend of Schöner's. He's very good. The best printer of scientific works anywhere in Germany. In all of Europe, probably. If I showed him this manuscript . . .

COPERNICUS. I told you, I've decided not to publish.

RHETICUS. You can't keep this to yourself. It isn't right. Secrecy has no place in science anymore.

COPERNICUS. Easy for you to say. You would not face the scorn that I have to fear.

RHETICUS. The mathematicians will . . .

COPERNICUS. Not just mathematicians but Church men will oppose me.

RHETICUS. After you publish it, if someone disagrees with you . . .

COPERNICUS. *If* someone disagrees? If?!

RHETICUS. If someone disagrees with you, let him publish a counterargument. Then you come back to refute his counterargument. And you go on like that. Back and forth. That's how learned men make good use of the God-given printing press.

COPERNICUS. It . . . it isn't even finished.

RHETICUS. There's enough material here to . . .

COPERNICUS. No. Several sections still need work.

RHETICUS. Show me.

ANNA *enters.*

ANNA. His clothes are dry.

RHETICUS. (*to* COPERNICUS) Let me help you.

ANNA. Everything's ready. It's time now.

COPERNICUS. Thank you, Anna. You can just leave those here.
ANNA. He should be going soon.
COPERNICUS. Leave the things, Anna. I'll be down shortly.
ANNA. You know he doesn't belong here.
COPERNICUS. I said . . .
ANNA. He could ruin everything!
COPERNICUS. Anna, don't . . .
ANNA. Why are you doing this?! What is the matter with you?!

ANNA, *crying, runs out of the room.*

COPERNICUS. (*following her out*) Anna, wait!

RHETICUS *picks up his clothes, dresses, though still unable to tear himself away from the manuscript. He continues to devour it, turning toward the door a couple of times with an uneasy sense of being watched. After another few moments, he sidles to the door, opens it.*

FRANZ, *who had been kneeling at the keyhole, falls into the room.*

RHETICUS. Hello.
FRANZ. (*getting up*) The bishop was concerned . . .
RHETICUS. The bishop?
FRANZ. About . . . your health.
RHETICUS. Am I under arrest?
FRANZ. Oh, no. Not yet. I mean . . . I don't think so, no. But I'm supposed to watch you.
RHETICUS. Were you watching me just now?
FRANZ. No.
RHETICUS. You must have seen me . . .
FRANZ. No, I didn't see anything . . .
RHETICUS. You must not have been looking very carefully.
FRANZ. Are you . . . feeling better?
RHETICUS. That depends. Who wants to know?
FRANZ. I do.
RHETICUS. Fine, thank you. Very much improved. You may call me Joachim.

FRANZ *stares at him, too flustered to speak.*

RHETICUS. And you are?

FRANZ. What are all these things? What does he do up here?

RHETICUS. He makes the Earth move.

Beat.

RHETICUS. Come here. I'll show you. Don't be afraid.

RHETICUS *takes* FRANZ *by the shoulders, repeating the earlier action of* COPERNICUS, *but gently.*

RHETICUS. He says the Earth turns, you see. It sounds silly, the first time you hear it, I know, but it gives you a way to explain why you see the Sun come up every day, then slowly move . . .

RHETICUS *turns* FRANZ, *step by step.*

RHETICUS. Across the sky, until it sets in the west.

RHETICUS *stops Franz with his back to him, moves closer to him.*

RHETICUS. Then it's nighttime. But the Earth doesn't stop there. It keeps on turning, through the night.

RHETICUS *continues turning* FRANZ *slowly, to face him.*

RHETICUS. Until dawn ends the darkness, and it's day again.

RHETICUS *leans closer, takes* FRANZ'S *face in his hands.*

Lights fade to black.

SCENE X. BISHOP'S BEDROOM
SCRIPTURE

COPERNICUS *conducts a routine, hands-on medical exam of the* BISHOP *as they speak.*

BISHOP. Has he come round, then?

COPERNICUS. He has, yes.

BISHOP. What have you learned about him?

COPERNICUS. It's rather something he taught me, about myself, Your Reverence.

BISHOP. Oh, yes? But who is he?

COPERNICUS. He has awakened in me a wish to resume my own work in mathematics.

BISHOP. Is that so?

COPERNICUS. The explication of my theory.

BISHOP. I don't know, Nicholas. Do you really have time for that? With all your other duties?

COPERNICUS. I feel I must make time for it, now.

BISHOP. But how? When? You're not a young man, you know.

COPERNICUS. Precisely.

Beat.

BISHOP. You still haven't told me anything about him.

COPERNICUS. I'm thinking I should publish it.

BISHOP. Publish? Your theory?

COPERNICUS. He has convinced me that other mathematicians will welcome the idea.

BISHOP. Your idea, Nicholas . . . Well. I'm sure it's very mathematical. Indeed. But, at the same time, it oversteps the bounds of mathematics. As I see it. I will go so far as to say it shakes the very foundation of Church doctrine.

COPERNICUS. Oh, no, Your Reverence.

BISHOP. What about Joshua?

COPERNICUS *sighs.*

BISHOP. Answer me.

COPERNICUS. Begging Your Reverence's pardon. It's just that I . . . I have had Joshua raised against me so many times that I begin to feel myself like one of his enemies among the Amorites.

BISHOP. Well, what about him?

COPERNICUS. Nothing.

BISHOP. How do you respond, Nicholas? How do you defend yourself against the charge that your ideas conflict with what the Bible says of Joshua?

COPERNICUS. I don't answer. I think it's better to say nothing.

BISHOP. You refuse to answer me?

COPERNICUS. Oh, not you, Your Reverence. I don't want to answer the charge. I would rather avoid any mention of Joshua, and limit my comments to mathematics alone.

BISHOP. That's no answer.

COPERNICUS. I'm afraid, Your Reverence. Afraid there may be . . . babblers, who claim to be judges of astronomy, although completely ignorant of the subject. And such men are not above twisting some passage of Scripture to their purpose, to censure me.

BISHOP. I am not trying to censure . . .

COPERNICUS. Oh, I know Your Reverence is not one of those.

BISHOP. I have overlooked all sorts of infractions lately, as I need not remind you!

COPERNICUS. Thank you, Your Reverence.

BISHOP. But you will most certainly have to deal with Joshua. And countless other passages of Holy Writ. The Psalms also teach us that the Earth does not move.

COPERNICUS. As I read those passages, I hear the godly Psalmist declare that he is made glad through the work of the Lord. That he rejoices in the works of His hands. Only that.

BISHOP. Are we reading the same Bible, Nicholas? Psalm 104 says the Lord God laid the foundation of the Earth, that it not be moved forever. Not be moved. Forever.

COPERNICUS. It's a matter of interpretation.

BISHOP. What's to interpret? It's stated there in plain language. It couldn't be more clear. Not to be moved forever. It doesn't say it should spin like a top.

COPERNICUS. To me, it says that God, the source of all goodness, created an abiding home for mankind on this Earth. And that foundation will hold firm forever.

BISHOP. That still doesn't answer Joshua.

COPERNICUS. As strange as it may sound, Your Reverence, to someone who is not a mathematician, my theory offers certain advantages for the improvement of the calendar.

BISHOP. The ecclesiastical calendar?

COPERNICUS. Easter, for example. To calculate the correct date of Easter each year.

BISHOP. You could make a contribution of that significance?

COPERNICUS. I don't mean to boast.

BISHOP. Why didn't you say so before? Why have you never even mentioned the calendar until now?

COPERNICUS. I lacked the confidence to expose my theory . . . to the scrutiny of others.

BISHOP. I had no idea there was so much to it.

COPERNICUS. Then . . .

BISHOP. All these years, I thought you were just . . .

COPERNICUS. Do I have Your Reverence's blessing to continue the work?

BISHOP. I suppose so. If what you say is true, then I suppose you *should* take it up again.

COPERNICUS. Thank you, Your Reverence.

BISHOP. I had no idea. How does that work, then? How does what you do relate to the date of Easter?

COPERNICUS. It concerns correcting the exact duration of the tropical year, from equinoctial and solstitial observations of . . .

BISHOP. Never mind that now.

COPERNICUS. I'm sure I could explain . . .

BISHOP. Yes, yes. So now I suppose you'll need to find a press. A printer.

COPERNICUS. I hear there is an excellent one in Germany.

BISHOP. Tosh! Have we no printers here in Poland?

COPERNICUS. Mmm. None, I think, that could take on a work of this nature.

BISHOP. Such a book could bring very positive attention to Varmia. Not just Varmia. To Poland. To . . . It should definitely be printed here.

COPERNICUS. It's such a lengthy work . . .

BISHOP. Who do you know in Germany?

COPERNICUS. For this kind of text, with the large number of diagrams required . . .

BISHOP. Aren't you getting ahead of yourself? Don't you think you should write this book before you worry about where to have it printed?

COPERNICUS. Yes, Your Reverence. There is much to be done.

BISHOP. So it's a big book, is it?

COPERNICUS. I've already written several hundred pages, over the years.

BISHOP. As much as that?! My, my. And still not come to the end?

COPERNICUS. It's . . . complex.

BISHOP. You know what I'm thinking, Nicholas? I'm thinking you'll want to leaven your mathematical jargon with a little poetry. What do you think?

COPERNICUS. I hadn't given it any thought.

BISHOP. Your book will need that literary touch. When the time comes, I shall compose verses for you to include, as an introduction. An invocation.

COPERNICUS. Your Reverence is too kind.

BISHOP. A scholarly book like this will surely attract the duke's attention. He'll see what talents we have here in . . . Why, the king himself might recognize the diocese for such a . . . How long do you think it will take you?

COPERNICUS. That is difficult to say. I'll need help to complete the project.

BISHOP. You may have my secretary. I'll put him at your disposal.

COPERNICUS. That's not the kind of help that I . . .

BISHOP. Ah! You mean the mathematics part.

COPERNICUS. Yes, Your Reverence.

BISHOP. There's no one fit to assist you with that.

COPERNICUS. I suspect, Your Reverence, that the unfortunate invalid who recently fell ill at my door might be . . .

BISHOP. Aha! Back to him again.

COPERNICUS. If he were amenable. And if Your Reverence will allow, I might ask him to stay on a while, as a collaborator.

BISHOP. Since when have you sought my approval on your houseguests, Nicholas?

COPERNICUS. This is a special case.

BISHOP. And the harlot? I suppose she, too, is crucial to your new endeavor?

COPERNICUS. He is Professor of Mathematics at Wittenberg.

The BISHOP *gasps.*

COPERNICUS. I did not invite him here. As God is my witness, I had no idea he was coming. He materialized on my doorstep like a . . . like . . .

BISHOP. How you insult me! Abuse me!

COPERNICUS. It was all such a coincidence that it seemed it could not be merely coincidence . . . As though there were something . . . providential in his arrival.

BISHOP. Silence! You think Heaven has sent you a Lutheran to help you tell the world your crazy idea?

Beat.

BISHOP. Damn it all, Nicholas! Do you expect me to break my own law to accommodate you?

COPERNICUS. No one need know who he is. I promise to keep him out of sight. His presence will offend not a soul. I swear it.

BISHOP. You can't have them both. Your choice, Nicholas. The harlot. Or the heretic. One or the other.

Blackout.

SCENE xi. TOWER ROOM
INFATUATION

The stars of the planetarium effect appear, start to spin, speeding up quickly.

FRANZ. Wheeeeeeeee!

RHETICUS. Is that fun?

FRANZ. Spin faster! Ooooooooooooh!

The lights slow to a stop.

FRANZ. Awwww.

RHETICUS. All right. One more time. Here we go.

Lights speed up, slow down, stop.

Stage lights return as RHETICUS *opens the hatch and* FRANZ *pokes his head out of the Machine.*

FRANZ. I still don't understand it. But I love it. I love how it looks when it . . .

RHETICUS. Come on out now.

FRANZ. Can't I go again?

RHETICUS. No, no. I have to get back to work.

RHETICUS *pulls him out, kisses him.*

RHETICUS. (*patting* FRANZ'S *bottom*) Run along now. But come back later.

FRANZ. I don't have to run along.

RHETICUS. I mean it. I promised him I'd get through the superior planets before he . . .

FRANZ. And I promised His Reverence the bishop to observe you closely. I'm to report everything you do. Every little thing. So you see, I am at my post, doing my duty. I don't have to run along anywhere.

They kiss again, embrace, make their way to the cot.

Blackout.

SCENE xii. COPERNICUS'S BEDROOM
PARTING

ANNA *stands stunned, pained.*

COPERNICUS. It's only for a little while.

ANNA. Only . . . ?

COPERNICUS. Just until he calms down. About the attempt on his life. You understand.

ANNA. But there was no attempt on his life. You said so yourself, he merely ate something rotten.

COPERNICUS. You know how irascible he is.

ANNA. How long will we be apart?

COPERNICUS. I don't know. But, I think, the sooner we acquiesce, the faster the whole business will settle. And then we can be together.

ANNA. Promise?

COPERNICUS. You'll see.

ANNA. But where am I supposed to go?

COPERNICUS. Bishop Giese has offered his . . .

ANNA. You have a place for me? Already?

COPERNICUS. It's just that he offered. To help us.

ANNA. So. It's all arranged. Everything settled.

COPERNICUS. I wouldn't add that to your burden. Of course I looked into lodgings for you. Temporary lodgings.

ANNA. I don't need new lodgings. This is where I live.

COPERNICUS. Anna . . .

ANNA. I have a right to be here.

COPERNICUS. I know that.

ANNA. What about him? He's the one who should be run out of town.

COPERNICUS. That's different. He's ill.

ANNA. Maybe I should get sick.

COPERNICUS. Anna.

ANNA. Why shouldn't I get sick? Then you'd have to take care of me, too.

COPERNICUS. No, Anna.

ANNA. Why not? Why couldn't he have infected me? I bathed him, touched his clothing. And now, you see, now I, too, am too weak to travel.

COPERNICUS. No.

ANNA. Why not, Mikoj? You could have your own little hospital up there. With two patients instead of one. And that way, we could . . . That way, I wouldn't have to be away from you a single night!

COPERNICUS. The bishop would see through that.

ANNA. That dog of a false priest. I see how he looks at me. That's why he's forcing you to send me away. That's the real reason, the old lecher.

COPERNICUS. Has he touched you?

ANNA. I can read what's on his mind, plain enough. But if I lie low, stay out of sight . . .

COPERNICUS. Let's not make this any more difficult than it is.

ANNA. Soon you'll have to send your professor away, you know. How long do you think you can pretend that he's just lying up there, unconscious? The bishop will find out. He has his spies, you know. Someone's bound to see that you're not tending to his fever when you go up there. What is it, Mikoj? Why are you looking at me like that? Surely you don't think I would give you away? Oh, Mikoj, I'll never tell a soul. You know I would never say or do anything that could hurt you.

COPERNICUS. I know that. I just wish it didn't have to be this way.

They embrace, cling to each other.

ANNA. Come with me.

COPERNICUS. What?

ANNA. Come away with me. Let's both go. Why should you grovel to him any longer?

COPERNICUS. You think I could simply walk away from here?

ANNA. Run away. Come and be with me where no one will care that we're together. Leave all these nasty, nosy old men.

COPERNICUS. Leave the Church?

ANNA. We'll make a new life. Our own life. Anywhere but here. Think of it, Mikoj. You could have a real hospital. I could be a midwife. We'll get by. You'll see.

COPERNICUS. I'm too old to change, Anna.

ANNA. Not so old.

COPERNICUS. We both knew, all along, that the two of us . . . that we could never have a life together.

ANNA. You won't go with me?

COPERNICUS. I can't.

ANNA. You can't.

COPERNICUS. I'm sorry.

ANNA. You can't. You! You turned the whole universe inside out and upside down. You told every planet which way to go. Are you still that man, Mikoj?

Blackout.

SCENE xiii. TOWER ROOM DEDICATION

Several weeks later. The room looks more "lived in." The manuscript has grown to several stacks of pages, more or less neatly arranged.

RHETICUS. I still say you make the case too quickly. You've got to work up to it. Broach the idea slowly.

COPERNICUS. I don't want to pretend the book is something it's not.

RHETICUS. You can't just push the Sun to the center of the universe on page one.

COPERNICUS. That's the whole point.

RHETICUS. Yes, but you've still got to build up to it, the way I tried to show you. You can't just pluck the lantern of the universe, for God's sake, from its place in the eternal, perfect heavens, and shove it into the Hell hole at the bottom of the world.

COPERNICUS. Later on, I explain why . . .

RHETICUS. Move the whole thing later. They'll all turn against you if you don't. They'll be clinging for dear life to the old, immobile Earth. They'll insist that the Earth belongs at the center because . . . because of its Earthiness. Because of all the change, and death, and decay. If you want to put the Sun there, in the midst of all that, you had better do it slowly.
Beat.

COPERNICUS. You mean I haven't proven it. Mathematically.

RHETICUS. I didn't say that.

COPERNICUS. But that's what you mean. If the proofs were stronger, you wouldn't be trying so hard to make it sound palatable.

RHETICUS. I want them to hear you out, to see what you've done. I'm begging you: Invite them into this new world. Don't foist it on them.

COPERNICUS. Maybe it isn't ready after all. Maybe this was all a big mistake.

RHETICUS. No, no. Don't say that.

COPERNICUS. I don't know what made me think I could . . .

RHETICUS. (*going to him, taking his shoulders to encourage him*) You mustn't lose heart. You've got to leave a few stones unturned. Something for others who come after you to do. You've given us so much to build on. Your work is . . . It's like that cathedral out there. Do you think anyone who laid the stones for the foundation was still around when the cross went up on top? Trust me, Father. A hundred years from now, astronomers will still be reading your book.

COPERNICUS. And you, Joachim?

RHETICUS. I will have read it a hundred times.

COPERNICUS. What will you do after we finish here?

RHETICUS. After? I will take your book to Nuremberg. I'll watch over the printer, to keep him on his toes. I'll proofread every page, I'll . . .

COPERNICUS. *After* that.

RHETICUS. I don't have to worry about anything after that.

COPERNICUS. You'll go back to Wittenberg? To your teaching?

RHETICUS. No, Father. By that time . . . By then I'll be . . .

COPERNICUS. What?

RHETICUS. There is no "after" for me, after that. Don't you remember? By this time next year, when Jupiter and Saturn enter their Great Conjunction, my time will be . . .

COPERNICUS. You can't still believe that.

RHETICUS. Nothing in your theory gives me a way out.

COPERNICUS. You can't just resign from life. Acquiesce to some benighted . . .

RHETICUS. I have accomplished my mission. That's something. Not many old men can say as much. I found you. I pulled your work out of the rubbish heap. And once I see it published, I'm done. It will no longer matter what happens to me.

COPERNICUS. You don't know what will happen.

RHETICUS. But I do.

COPERNICUS. You could live a hundred years. You have no idea what the future holds for you.

RHETICUS. You've done everything you could do for me. The time with you has been . . .

COPERNICUS. Wait and see what happens to your career when Schöner and the rest of them read my acknowledgments to you.

RHETICUS. To me?

COPERNICUS. Of course to you.

RHETICUS. Oh, no. You mustn't disclose my role in this.

COPERNICUS. You think I wouldn't thank you, publicly, for all you did to . . .

RHETICUS. My name must not appear in your book. It would taint the whole thing.

COPERNICUS. I don't care. I owe you . . .

RHETICUS. No. You have others you can thank, without inflaming the leaders of your Church.

COPERNICUS. Even the bishop knows how much you have . . .

RHETICUS. It's not the bishop I'm worried about.

COPERNICUS. Luther?

RHETICUS. I have a new plan, for a dedication that you will write. To the real power.

COPERNICUS. You mean Duke Albert?

RHETICUS. No!

COPERNICUS. The king?

RHETICUS. No, no one from the government. The dedication must acknowledge higher powers. Someone in the Church.

COPERNICUS. Not the bishop?

RHETICUS. No! It's bad enough we're stuck with his doggerel verses.

COPERNICUS. Tiedemann?

RHETICUS. Not nearly powerful enough.

COPERNICUS. Who then, the pope?

RHETICUS. Yes!

COPERNICUS. I was joking, Joachim.

RHETICUS. I am perfectly serious.

Beat.

RHETICUS. He's really the only one.

COPERNICUS. His Holiness?

RHETICUS. Paul Pontifex Maximus himself. To protect you. From those backbiters who will bend chapter and verse to evil purposes, and try to condemn your theory. Even though, we both know, there is nothing irreverent in your book, nevertheless there is the danger that someone will . . .

COPERNICUS. But . . . His Holiness.

RHETICUS. The mere mention of his name will lend the book the air of papal authority. It might even give people the impression that he had commissioned you to write it.

COPERNICUS. He would never do that.

RHETICUS. Still, it might appear that he had.

COPERNICUS. What could he possibly have to say about astronomy?

RHETICUS. He doesn't have to say anything. You simply dedicate the book to him.

COPERNICUS. I couldn't even do that without his express permission.

RHETICUS. Then we must get his permission.

COPERNICUS. He has the troubles of the world on his shoulders. He's gone and excommunicated the King of England.

RHETICUS. Your bishop must have representatives in Rome. Ambassadors to the Vatican? Someone who can get to him?

COPERNICUS. Even if we could get to him . . . He is consumed with a final solution to the Lutheran problem! I'm sorry, Joachim. Forgive me for . . .

RHETICUS. I have no love for him either. To me, he's the Antichrist. But for your book . . . Trust me, Father. If you dedicate your studies to him, then you prove to everyone that you do not run away from judgment, even by the highest authority.

COPERNICUS *considers this, smiles, then laughs. It's the first good laugh he's had in a long time, and he enjoys it.*

RHETICUS, *not sure of the joke, nevertheless joins in the laughter.*

COPERNICUS *hugs him, thumps his back, finally recovers enough breath to speak.*

COPERNICUS. I'm just picturing the bishop's face when I ask him to . . .

They both dissolve again. COPERNICUS *gives* RHETICUS *a fatherly hug and goes to the door. They share one more laugh, nodding at each other, serious again.*

Blackout.

SCENE xiv. BISHOP'S PARLOR
HERETICUS

BISHOP. That's all he does?

FRANZ. Yes, Your Reverence.

BISHOP. Just . . . writes?

FRANZ. Sometimes he walks around, thinking. Often the doctor is there, too, and they talk. But most of the time he writes.

BISHOP. No one else comes to the room?

FRANZ. No, Your Reverence.

BISHOP. No messages from . . . anywhere?

FRANZ. Not that I have observed, Your Reverence.

BISHOP. And no sign of the . . . the, uh, the housekeeper.

FRANZ *looks down, shakes his head no.*

BISHOP. Very well. You needn't watch him quite so closely any longer.

FRANZ. No?

BISHOP. It's time you got back to some of the tasks you've neglected. All right, then. You may tell the doctor I will see him now.

FRANZ *exits.*

COPERNICUS *enters.*

BISHOP. Come in, Nicholas. How are you getting along, you and Professor . . . Professor Hereticus?

COPERNICUS. Rheticus, Your Reverence. His name is Rheticus. And he is most grateful to Your Reverence for tolerating his presence all this time.

BISHOP. Don't tell me you need another extension on the time?

COPERNICUS. No. I'm here to report that our work is very nearly finished. Just a few more sections, and then . . .

BISHOP. Excellent!

COPERNICUS. Yes. Well. There is one more thing that we think might require Your Reverence's assistance to . . .

BISHOP. You see? It's what I told you all along. Now you have proved it to yourself. How getting rid of that hussy has freed your mind for the serious work God intended you to do.

Blackout.

SCENE XV. TOWER ROOM
DISCOVERY

FRANZ *and* RHETICUS *lie together on the cot in an embrace.*

FRANZ. (*getting up, starting to dress*) I have to get back.

RHETICUS. What about your duty here? To me?

FRANZ. He's finding a million other things for me to do now. Every day it's something new.

RHETICUS. (*getting up, going to him*) What about the nights?

They kiss.

RHETICUS. Come back later. Promise?

They kiss again.

Copernicus enters, sees them, and reels.

They see him as well. FRANZ *jumps, starts to bolt, but* RHETICUS *holds him.*

FRANZ *breaks free, runs out.*

RHETICUS. You've known all along. Haven't you?

COPERNICUS. I wasn't sure.

RHETICUS. But you suspected?

COPERNICUS. I prayed that my suspicions were unfounded.

RHETICUS. Now you know the truth.

COPERNICUS. Yes.

RHETICUS. And you despise me.

COPERNICUS. No, Joachim. Neither do I judge you.

RHETICUS. You needn't pretend to understand.

COPERNICUS. But I can no longer protect you.

RHETICUS. From myself?

COPERNICUS. Do you know what will happen to you, if you are discovered?

RHETICUS. I know.

COPERNICUS. You couldn't know, or you wouldn't . . .

RHETICUS. I know!

COPERNICUS. The law condemns anyone who commits . . .

RHETICUS. Don't quote me the law.

COPERNICUS. It says you forfeit your life.

RHETICUS. It doesn't matter.

COPERNICUS. You will be burned alive!

RHETICUS. Burn alive and die! Die and burn in Hell forever! I am doomed either way.

COPERNICUS. If you're discovered . . . If word of this should reach the boy's father . . .

RHETICUS. He won't dare tell his father. He won't tell anyone.

COPERNICUS. The risk is too great, Joachim.

RHETICUS. He won't tell.

COPERNICUS. You've got to get out of here. Go now, before anything else happens.

RHETICUS. Go?

COPERNICUS. Go! Yes! Now. I insist that you go.

RHETICUS. I can't abandon you now.

COPERNICUS. I won't have you risk your life for the sake of . . .

RHETICUS. I don't care what happens to me.

COPERNICUS. Then think of the boy. Don't ruin his chances . . .

RHETICUS. We're so close to the end. Another few days is all we . . .

COPERNICUS. No.

RHETICUS. Just . . .

COPERNICUS. It's impossible. Not another word now. Off with you, or I'll die of fright.

RHETICUS. Let me finish what we . . .

COPERNICUS. I'm afraid for you, Joachim.

RHETICUS. All right, I'll go.

COPERNICUS *puts his hand to his heart, sits down.*

RHETICUS *begins gathering the piles of manuscript pages.*

COPERNICUS. What are you doing?

RHETICUS. I'll take it to Nuremberg. Do what I promised.

COPERNICUS. No. You can't . . .

RHETICUS. To the printer.

COPERNICUS. No.

RHETICUS. I will keep that promise, no matter what.

RHETICUS *continues packing the manuscript.*

COPERNICUS *tries to stop him, grabs the pages from him.*

COPERNICUS. Stop!

RHETICUS. (*refusing to let them go*) What's the matter with you?

COPERNICUS. It's not ready.

RHETICUS. It is.

COPERNICUS. No.

RHETICUS. It's . . .

COPERNICUS. I'm not ready.

RHETICUS. I'll take this much with me now, and later you can send . . .

COPERNICUS. You cannot take my manuscript!

RHETICUS. Have you completely lost faith in me?

COPERNICUS. No.

RHETICUS. I will guard it with my life.

COPERNICUS. No.

RHETICUS. You know I will.

COPERNICUS. No.

RHETICUS. I swear it.

COPERNICUS. I never meant for you to take it away.

RHETICUS. We've been working toward this moment ever since we . . .

COPERNICUS. I need to keep it here. With me.

RHETICUS. I promised you a published book.

COPERNICUS. Keep it by me.

RHETICUS. I've got to take it with me to . . .

COPERNICUS. Not this. No.

RHETICUS. But how can I . . . ?

COPERNICUS. A copy. I meant for you to take a copy with you. Not my manuscript.

RHETICUS. I've copied only the first few chapters. Not enough to . . .

COPERNICUS. I can't let it go.

RHETICUS. There isn't time now to copy . . .

COPERNICUS. (*breaking down*) I can't.

RHETICUS. You've got to let me . . .

COPERNICUS. I can't. It's been with me my whole life. It is my life. I cannot part with it.

COPERNICUS *clutches the manuscript to his chest.*

RHETICUS. All right. I'll just take the parts I copied.

COPERNICUS. I can't.

RHETICUS. It's all right. But what about the rest? How will you . . . ?

COPERNICUS. I'll copy it for you. I'll . . .

RHETICUS. You can't do that by yourself.

COPERNICUS. I'll get someone to help me. You'll see.
RHETICUS. I'll be waiting for it. In Nuremberg.
COPERNICUS. I know you will.
RHETICUS. I won't fail you.
COPERNICUS. Go now.
RHETICUS. Everyone will be waiting for it.
COPERNICUS. Yes. Now go.

RHETICUS *closes the satchel, looks around the room.*

COPERNICUS. Joachim!

COPERNICUS *embraces* RHETICUS *in a long good-bye hold.*

COPERNICUS. Good-bye, Joachim.
RHETICUS. Good-bye, my teacher.

RHETICUS *goes to the door, turns for a last look.*

COPERNICUS. May God forgive you, and bless you.
RHETICUS. God be with you, my teacher. My father.

RHETICUS *exits.*

COPERNICUS. And with you. May God be w- w- w-

COPERNICUS *shakes his head to clear it, tries to speak. His right arm falls to his side, but he still clutches the manuscript to his chest with the left as he sinks into a chair.*

Blackout.

SCENE XVI. COPERNICUS'S BEDROOM
DE REV

COPERNICUS *lies in bed, comatose.* GIESE *kneels beside him, praying. A loud knocking comes from the front door, but* GIESE *tries to ignore it.*

ANNA. (*offstage*) For pity's sake! Let me in. Oh, why won't you open the door? Let me in, I say. Have you no heart?

GIESE *relents, goes to the door.*

ANNA. (*offstage*) Let me in. Let me in!

GIESE *opens the door.*

ANNA. (*entering*) Why didn't you tell me? Why didn't anyone say anything? Oh, where is he?

ANNA *rushes past* GIESE *to the bedside.*

GIESE *stays close behind her.*

ANNA. Oh, Mikoj! It's me, my dearest. I'm here with you now. It's all right. They didn't want me to know, but I found out. And now I'll stay with you. I'll be here every minute. Don't worry. I'm here.

GIESE. He doesn't hear you.

ANNA. Shh. Look! He's trying to speak.

GIESE. He hasn't said a word in weeks now. Nothing.

ANNA. But his eyes are open. His lips are moving. Look.

GIESE. The duke sent his personal physician. He said that's just a . . . a reflex.

ANNA. You don't know that. He may hear everything we're saying. (*to* COPERNICUS) Can you hear me, Mikoj? You don't have to talk if you don't want to. If it's too hard for you, you just rest. I know. It's all right. I'm not leaving you now.

GIESE. There's nothing to be done.

ANNA. You should have told me.

GIESE *puts a hand on her head, as though to bless her, but she stands up to face him.*

ANNA. (*whispering*) He wouldn't want this.

GIESE. He is not afraid to die.

ANNA. There are certain powders I know about. They could . . . end his suffering.

GIESE. God will take him when it's time.

ANNA. I'm just saying, it would be possible to ease his . . . Even to hasten his entry into the next life.

GIESE. You mustn't say such things, my child. You must not even think them.

ANNA *kneels by the bed again, takes* COPERNICUS'S *hand.*

GIESE *prays.*

FRANZ. (*running in*) Bishop Giese! It's here, Bishop Giese! It's here! Look!

GIESE. Hush. What's . . . ?

FRANZ. Miss Anna!

GIESE. What have you got there?

FRANZ. It's from Nuremberg. See? This must be it.

GIESE. Let me look.

ANNA. Something for him?

GIESE. Let's just see what we have here.

FRANZ. Is it . . . ?

GIESE. Look at that!

ANNA. What is it?

FRANZ. I knew it!

ANNA. Is that his book?

FRANZ. All those hundreds of pages I copied for him. For both of them.

GIESE. I never thought I'd see the day.

FRANZ. Is there a note? Any word from . . . ?

ANNA. That can't be it. Just a pile of paper?

GIESE. "On the Revolutions of the . . ."

ANNA. That's really it?

GIESE. "Heavenly Spheres." By . . .

FRANZ. There's nothing else in the package?

GIESE. Nicolaus Copernicus.

ANNA. It's not at all what I imagined. That professor played a mean trick on him.

FRANZ. No.

ANNA. How shabby it looks. That will never impress anyone.

GIESE. Oh, but it will. It will. This is just the way books come from the printer. Just the pages, like this. But I shall have it bound for him. Something very grand, in red leather, with his name stamped in gold letters. Wait till you see it then.

ANNA. Let's show it to him.

GIESE. All the times I urged him to do this . . . And how he fought against me. (*with a fond look at* COPERNICUS) The stubborn old mule.

ANNA. We should let him see his book.

FRANZ *takes a few sections and gives them to* ANNA.

ANNA *turns her full attention to* COPERNICUS, *showing him the book, ignoring the other two.*

GIESE *keeps looking at the rest.* FRANZ *peers over his shoulder.*

GIESE. I remember when he watched this eclipse. I went along with him to see it.

ANNA. It's here, Mikoj.

FRANZ. Where was that, Your Reverence?

ANNA. It's finally here.

GIESE. Right out there, in the meadow. The Moon was full. So bright. You could have read this book out there in the moonlight. It was that bright.

ANNA. (*propping him up*) I want you to have a good long look at this.

GIESE. I must have fallen asleep while waiting for it to start, because I remember how he woke me when it was time. He wouldn't leave the instruments, even for a moment, so he made a . . . a howling sound. Like a wolf! Awoooooow!

ANNA. It's your book, Mikoj. Your very own book, that you wrote.

GIESE. I jumped up. But then everything happened so slowly, very gradual. It took an hour, I think, or maybe more, for the shadow to completely cover the Moon.

ANNA. All your work, all those years, and here it is, at last.

GIESE. And you know what happened then? The Moon turned red.

FRANZ. Really?

ANNA. (*putting pages in his hands*) Hold it. Feel it. Isn't it wonderful?

GIESE. One of the most beautiful sights I ever saw.

ANNA. Mikoj?

COPERNICUS *slumps over, letting the pages fall to the floor.*

Blackout. Choir sings "Salve Mater Misericordiae."

SCENE xvii. CEMETERY FUNERAL

The bowed heads of mourners, with the BISHOP *presiding, suggest a graveside.*

BISHOP. Blessed are they that mourn, for they shall be comforted.

ALL. Amen.

BISHOP. Blessed are you, Father, Lord of Heaven and Earth, for revealing the mysteries of Thy Kingdom.

ALL. Amen.

BISHOP. I am the resurrection and the life. He that believeth in me, though he were dead, yet shall he live. And whosoever liveth and believeth in me shall never die.

ALL. Amen.

ANNA. (*pushing her way toward the* BISHOP) This, too!

BISHOP. Who is this woman?

ANNA. It's only right.

FRANZ. (*coming to her aid*) What are you doing, Miss Anna?

ANNA. He would want this. I know he would.

ANNA *starts to throw her package into the grave.*

BISHOP. Stop her.

FRANZ. (*taking the package*) Let me help you, Miss Anna.

ANNA *lets herself cry in* FRANZ'S *arms.*

GIESE. (*receiving the package from* FRANZ) It's his manuscript.

BISHOP. Get her out of here.

FRANZ. Come with me, Miss Anna.

FRANZ *walks* ANNA *downstage.*

The BISHOP *and* GIESE *remain behind, in darkness.*

RHETICUS *steps in front of* ANNA *and* FRANZ.

FRANZ. I knew you'd come back.

ANNA. You!

RHETICUS. (*to* FRANZ) Were you with him at the end?

ANNA. He doesn't need you anymore.

GIESE. (*joining them*) What seems to be the trouble here?

FRANZ. I said he'd come, didn't I? I knew he would.

ANNA. (*to* GIESE) Don't let him have that. He doesn't deserve it.

GIESE. (*to* FRANZ) Take her someplace where she can sit and rest.

FRANZ *obeys.*

ANNA. (*exiting, crying*) Let that go to the grave with him. He would want to have it with him.

GIESE. You recognize this, of course.

RHETICUS. Did he see it? The finished book?

GIESE. Oh, yes. It arrived just in time.

RHETICUS. The moment I left it with the courier, I thought, "Why did I do that? Why don't I go and give it to him myself?" But it had already gone. I started out the next morning, hoping to . . . And now . . .

GIESE. He was so pleased to see it. To hold it in his hands. Yes. And then he . . .

RHETICUS. But he did see it? He knew that I . . .

GIESE. He knew. Yes, my son. We are all so grateful to you, for what you've done. When I will read his book, it will bring him back to life for me.

GIESE *bows his head, grieving.* RHETICUS *also bows, puts a hand on* GIESE'S *shoulder.*

GIESE. (*handing him the manuscript*) Here. You should have this. As much as I would like to keep it for my own . . . for my comfort . . .

RHETICUS. He wouldn't let me take it.

GIESE. Now it belongs to you.

RHETICUS. You keep it. You're his . . .

GIESE. No. You have been the chief instigator in this affair. It's yours.

RHETICUS *takes the manuscript.*

GIESE. He'll never know what anyone thought about it . . . What people will say when they . . .

RHETICUS. No. They can say what they will, and he'll never know.

GIESE. What are they saying?

RHETICUS. I'm almost glad that he can't . . .

GIESE. What's the reaction? Do you know?

RHETICUS. It's . . . not as bad as he thought. Not what he feared.

GIESE. But not . . . good?

RHETICUS. No one is ready for what he had to say. The mathematicians I know, they're happy. They just take what they need from the book, and ignore the rest.

GIESE. Ignore it?

RHETICUS. They skip over that part.

GIESE. I didn't think anyone *could* ignore an idea like that.

Beat.

GIESE. But you believed him?

RHETICUS. He had no real proof.

GIESE. God rest his soul.

Beat.

GIESE. Everything is so still.

Beat.

RHETICUS. Is it?

GIESE. Hm?

RHETICUS. You know. Is it still? Or is it . . . ?

GIESE. What do you think?

RHETICUS. Sometimes, when I remember how he . . . When I hear his voice inside my head, I swear, I can almost feel it turning.

Blackout. The end.

PART THREE
Aftermath

One generation passeth away, and another generation cometh: but the Earth abideth for ever.
The Sun also ariseth, and the Sun goeth down, and hasteth to his place where he arose.
The wind . . . whirleth about continually, and the wind returneth again according to his circuits.
All the rivers run into the sea; yet the sea is not full; unto the place from whence the rivers come, thither they return again.
All things are full of labor; man cannot utter it; the eye is not satisfied with seeing, nor the ear filled with hearing.
The thing that hath been it is that which shall be; and that which is done is that which shall be done: and there is no new thing under the Sun.

—ECCLESIASTES 1:4–9

"A generation passes away, and a generation comes, but the Earth stands forever." Does it seem here as if Solomon wanted to argue with the astronomers? No; rather, he wanted to warn people of their own mutability, while the Earth, home of the human race, remains always the same, the motion of the Sun perpetually returns to the same place, the wind blows in a circle and returns to its starting point, rivers flow from their sources into the sea, and from the sea return to the sources, and finally, as these people perish, others are born. Life's tale is ever the same; there is nothing new under the Sun.

You do not hear any physical dogma here. The message is a moral one, concerning something self-evident and seen by all eyes but seldom pondered. Solomon therefore urges us to ponder.

—JOHANNES KEPLER, *Astronomia nova*, 1609
(TRANSLATED FROM THE LATIN BY WILLIAM H. DONAHUE)

CHAPTER 7
The First Account

It is also clearer than sunlight that the sphere which carries the Earth is rightly called the Great Sphere. If generals have received the surname "Great" on account of successful exploits in war or conquests of peoples, surely this circle deserved to have that august name applied to it. For almost alone it makes us share in the laws of the celestial state, corrects all the errors of the motions, and restores to its rank this most beautiful part of philosophy.

—GEORG JOACHIM RHETICUS, FROM THE *First Account*, 1540

O ONE KNOWS WHAT THE BRILLIANT, fervent young Rheticus said when he accosted the elderly, beleaguered Copernicus in Frauenburg. It is safe to assume he did not laugh at the idea of the Earth in motion. And maybe that was enough to make Copernicus open his long-shelved manuscript, and also his heart, to the visitor who became his only student. Rheticus's enthusiasm for astronomy leapt the barriers of age, outlook, and religious difference that might well have separated the two men. As Rheticus recalled years later of their time together, "Driven by youthful curiosity . . . I longed to enter as it were into the inner sanctum of the stars. Consequently, in the course of this research I sometimes became downright quarrelsome with that best and greatest of men, Copernicus. But still he would take delight in the honest desire of my mind, and with a gentle hand he continued to discipline and encourage me."

Nor does anyone know how Rheticus's presence in Frauenburg

escaped the wrath—or even the notice—of Bishop Dantiscus. Whether Copernicus intentionally hid the youth at first, or merely concealed his full identity, he soon hustled him out of town.

As Rheticus explained the situation later in a letter to a friend, "I had a slight illness, and, on the honorable invitation of the Most Reverend Tiedemann Giese, bishop of Kulm, I went with my teacher to Löbau and there rested from my studies for several weeks." Once outside Varmia, Rheticus was safe from religious persecution. The peace-loving Giese, who had long prodded Copernicus to publish his theory, must have been ecstatic to learn of their visitor's ties to a respected printer of scientific texts. For Rheticus had brought as gifts three volumes bound in white pigskin, containing an assemblage of five important astronomy titles, three of which had been set in type and ornamented by the eminent printer Johannes Petreius of Nuremberg.

By summer's end in 1539, Rheticus had learned enough from Copernicus to write an informed summary of his thesis. He framed this précis as a letter to another mentor, Johann Schöner, a widely respected astrologer, cartographer, and globe maker in Nuremberg—and presumably the person who referred Rheticus to Copernicus in the first place.

"To the illustrious Johann Schöner, as to his own revered father, G. Joachim Rheticus sends his greetings," the report began. "On May 14th I wrote you a letter from Posen in which I informed you that I had undertaken a journey to Prussia, and I promised to declare, as soon as I could, whether the actuality answered to report and to my own expectation." He then explained how his "illness" had diverted him to Kulm for a time. After ten weeks of concentration, however, he was ready to "set forth, as briefly and clearly as I can, the opinions of my teacher on the topics which I have studied."

Rheticus may have read a copy of the *Brief Sketch* in Schöner's

The erudite Johann Schöner of Nuremberg, as painted by Lucas Cranach the Elder.

library before visiting Copernicus, or he could have come with only a vague notion of the new cosmology. Now he found himself one of two or at most three people in the world to have paged through the complete draft version of *On the Revolutions*.

"My teacher has written a work of six books," he told Schöner, "in which, in imitation of Ptolemy, he has embraced the whole of astronomy, stating and proving individual propositions mathematically and by the geometrical method." Ticking off the topics covered in each of the six parts, Rheticus said nothing of what is today considered the work's most salient feature. Indeed, he remained strangely silent about the motion of the Earth until page nineteen of his protracted description. Perhaps he knew Schöner and other readers would find a moving Earth ludicrous, and so he avoided mention of it for as long as he could. Or, equally likely, he judged a different aspect of Copernicus's work to be more important, and therefore gave it precedence. This was the explanation of the eighth sphere, or how the daily spin of the heavens drifted slowly backward over time—the subject of Copernicus's spar with Werner. Rheticus presented Copernicus's numerical results without saying that the starry sphere remained stationary in the Copernican model. He concentrated instead on cyclic time patterns Copernicus had identified through observations of the Sun and stars. To Rheticus's mind, these long cycles coincided with turning points in world history, and he seized on an interpretation that he believed Schöner would appreciate:

"We see that all kingdoms have had their beginnings when the center of the eccentric [here Rheticus refers to long-term changes in the Sun's apparent position] was at some special point on the small circle. Thus, when the eccentricity of the Sun was at its maximum, the Roman government became a monarchy; as the eccentricity decreased, Rome too declined, as though aging, and then fell. When the eccentricity reached the boundary and quadrant of mean value, the Mohammedan faith was established; another great empire came into being and increased very rapidly, like the change in the eccentricity. A hundred years hence, when the eccentricity will be at its minimum, this empire too will complete its period. In our time it is at its pinnacle from which equally swiftly, God willing, it will fall with a mighty crash. We look forward to the coming of our Lord Jesus Christ when the center of the eccentric reaches the other boundary of mean value, for it was in that position at the creation of the world."

Surely Rheticus had found exactly what he had come for: Copernicus's carefully developed mathematical treatise offered a firm new footing for the most momentous predictions of astrology. Nothing could extend Rheticus's own longevity, of course, but in his short life he believed he might yet shape a destiny, and maybe even achieve glory, by bringing Copernicus out of the shadows.

"A boundless kingdom in astronomy has God granted to my learned teacher," Rheticus interrupted his narrative to exclaim. "May he, as its ruler, deign to govern, guard, and increase it, to the restoration of astronomic truth. Amen."

Next Rheticus boasted to Schöner how Copernicus had resolved the Moon's motion without stretching or shrinking the lunar diameter. One could talk easily enough about the Moon's going around the Earth without referring to the Earth's motion. Only when he brought up the motions of the other planets did Rheticus finally concede that the center of the universe might shift in the new picture. And, almost in the same breath, he defended the switch: "Indeed, there is something divine in the circumstance that a sure understanding of celestial phenomena must depend on the regular and uniform motions of the terrestrial globe alone."

The road ahead—meaning the struggle to convince others to accept Copernicus's wisdom—would admittedly be difficult. But Rheticus had committed himself to this course and expected Schöner to do the same.

"Hence you agree, I feel, that the results to which the observations and the evidence of heaven itself lead us again and again must be accepted, and that every difficulty must be faced and overcome with God as our guide and mathematics and tireless study as our companions." Even Ptolemy, Rheticus claimed, "were he permitted to return to life," would applaud this "sound science of celestial phenomena."

Rheticus whipped himself into a frenzy of enthusiasm as he appraised Copernicus's labors. He found it almost inconceivable to contemplate the burden of effort that had allowed his teacher to take all the disparate phenomena of astronomy and link them "most nobly together, as by a golden chain." Over the remainder of his sixty-six-page report (twice as long as the *Brief Sketch* and the *Letter Against Werner* combined), Rheticus directly addressed Schöner a dozen times, as though to shake him awake to the new reality: "To offer you some taste of this matter, most learned Schöner," "Let me in passing call your attention, most learned Schöner," "But that you may the more readily grasp all these ideas, most learned Schöner," and so on, leading up to a final impassioned appeal:

"Most illustrious and most learned Schöner, whom I shall always revere like a father, it now remains for you to receive this work of mine, such as it is, kindly and favorably. For although I am not unaware what burden my shoulders can carry and what burden they refuse to carry, nevertheless your unparalleled and, so to say, paternal affection for me has impelled me to enter this heaven not at all fearfully and to report everything to you to the best of my ability. May Almighty and Most Merciful God, I pray, deem my venture worthy of turning out well, and may He enable me to conduct the work I have undertaken along the right road to the proposed goal."

The letter, had it been merely a letter, might have ended there. But Giese, together with Copernicus, hoped to see Rheticus's report published as a test of acceptability for the heliocentric theory, and therefore the ending necessarily took a political turn. In his effusive concluding pages, Rheticus sang the praises of Prussia.

"You might say that the buildings and the fortifications are palaces and shrines of Apollo; that the gardens, the fields, and the entire region are the delight of Venus, so that it could be called, not undeservedly, Rhodes. What is more, Prussia is the daughter of Venus, as is clear if you examine either the fertility of the soil or the beauty and charm of the whole land."

Rheticus extolled the Prussian forests teeming with stag, doe, bear, boar, aurochs, elk, and bison; the beehives, orchards, and plains; the rabbit warrens and birdhouses; and the lakes, ponds, and springs that he deemed "the fisheries of the gods." He cited the famous personages of the land as well, bowing respectfully to "the illustrious prince, Albrecht, Duke of Prussia" and that "eloquent and wise Bishop, the Most Reverend Johann Dantiscus."

Sometime between mid-May 1539, when Rheticus arrived in Varmia, and September 23, the date he concluded his report, Bishop Dantiscus likely discovered his presence through his network of informants. But he took no legal action. Perhaps Giese convinced Dantiscus of the visiting professor's value in publicizing Canon Copernicus's life work as a credit to Varmia. Or perhaps the Rheticus affair paled, in the bishop's eyes, in comparison to the continuing "trysts" between Copernicus and Anna Schilling. She had never left town, according to the gossip Dantiscus heard from his closest ally in Frauenburg, Provost Pawel Plotowski. Here Giese in fact interceded by letter, imploring Dantiscus not to believe such groundless rumors. Further aspersions from Plotowski, however, continued to inflame Dantiscus's anger.

"At his old age," Dantiscus complained of Copernicus in reply to Giese, "almost at the end of his allotted time, he is still said to receive his concubine frequently in furtive assignations." Dantiscus beseeched Giese to remonstrate with Copernicus on his behalf, and to speak as though he, Giese, were dispensing his own good advice. Reporting back to Dantiscus on September 12, 1539, Giese said he had reprimanded Copernicus as promised, but that his good friend denied all of Plotowski's pernicious charges.

Winding up the final pages of his report, Rheticus crafted an elaborate thank-you to Giese for his goodness and kindness, crediting the prelate with having served as Copernicus's inspiration.

"His Reverence mastered with complete devotion the set of virtues and doctrine, required of a bishop by Paul. He realized that it would be of no small importance to the glory of Christ if there existed a proper calendar of events in the Church and a correct theory and explanation of the motions. He did not cease urging my teacher, whose accomplishments and insight he had known for many years, to take up this problem, until he persuaded him to do so." This scenario, though otherwise undocumented, suggests that Giese became muse to Copernicus even before the heliocentric idea was born.

"Since my teacher was social by nature," Rheticus continued, "and saw that the scientific world also stood in need of an improvement of the motions, he readily yielded to the entreaties of his friend, the reverend prelate. He promised that he would draw up astronomical tables with new rules and that if his work had any value he would not keep it from the world. . . . But he had long been aware that in their own right the observations in a certain way required hypotheses which would overturn the ideas concerning the order of the motions and spheres that had hitherto been discussed and promulgated and that were commonly accepted and believed to be true; moreover, the required hypotheses would contradict our senses."

On the horns of that dilemma, Rheticus said, Copernicus had decided to "compose tables with accurate rules but no proofs." In other words, he would offer instructions for deriving planetary positions without mentioning his mind-boggling rationale. Rheticus doubtless knew from his weeks in their company that Giese and Copernicus had reached just such a compromise in 1535. Their friend Bernard Wapowski, recipient of the *Letter Against Werner,* visited them in Frauenburg in the autumn of that year, while Copernicus was completing an abbreviated treatise with a complete set of tables. Wapowski took a copy of this almanac with him back to Krakow. In October he tried, through his royal connections, to have the work printed in Vienna, but the negotiations ended in November, with Wapowski's death.

"Then His Reverence pointed out that such a work would be an incomplete gift to the world," Rheticus went on, "unless my teacher set forth the reasons for his tables and also included, in imitation of Ptolemy, the system or theory and the foundations and proofs upon which he relied." Thus Copernicus's book had come to be. Though the author later put it aside, Giese had never stopped agitating for its release.

"By these and many other contentions, as I learned from friends familiar with the entire affair, the learned prelate won from my teacher a promise to permit scholars and posterity to pass judgment on his work. For this reason men of good will and students of mathematics will be deeply grateful with me to His Reverence, the bishop of Kulm, for presenting this achievement to the world."

The other patron Rheticus esteemed loudly at the end of his report was Johann of Werden, the mayor of Danzig. "When he heard about my studies from certain friends, he did not disdain to greet me, undistinguished though I am, and to invite me to meet him before I left Prussia. When I so informed my teacher, he rejoiced for my sake and drew such a picture of the man that I realized I was being invited by Homer's Achilles, as it were. For besides his distinction in the arts of war and peace, with the favor of the muses he also cultivates music. By its sweet harmony he refreshes and inspires his spirit to undergo and to endure the burdens of office."

One would think Rheticus had laid on his hyperbole a bit thick, but his gushing got the *First Account* into the Danzig civic printing office, where it was published early in 1540. As soon as the first three sheets came off the press in March, a friend and classmate of Rheticus sent them to Philip Melanchthon in Wittenberg—evidence of what Rheticus, absent from the university for nearly two years now, had been doing with his time.

The title page of the *First Account* did not identify its author by name, but only as "a certain youth."

TO THE MOST ILLUSTRIOUS GENTLEMAN
MR. JOHANN SCHÖNER, CONCERNING
THE BOOKS OF THE REVOLUTIONS
Of the most learned Gentleman and
Most distinguished Mathematician,
The revered Doctor Mr. Nicolaus
Copernicus of Torun, Canon of
Varmia, By a certain youth
Most zealous for
Mathematics—
A FIRST ACCOUNT.

Rheticus, who had cleverly inserted his full name into the salutation in the *First Account*'s first paragraph, could afford to be modest on the title page. He sent copies of the finished book to friends and acquaintances who had helped him along the way, beginning naturally with Schöner.

Among the early plaudits to reach Rheticus in reply was a

congratulatory note from Andreas Osiander, the Lutheran theologian who had initiated Duke Albrecht's conversion. As other excited letters from scholars also greeted the *First Account*, Rheticus realized he was about to become famous. He could return to Saxony a hero.

From Giese's perspective, however, the publication of the *First Account* merely paved the way for *On the Revolutions*. He wanted the talented Rheticus to stay on in Frauenburg and help Copernicus prepare his lengthy manuscript for publication. Weary of finessing Rheticus's illegal residence in Varmia, Giese wished to find him a new sponsor—especially after April 15, 1540, when Bishop Dantiscus enforced the king's anti-Protestant decree by recalling all Varmian subjects "from the poisoned places of heretical Lutheranism," and calling also for the destruction of Lutheran books or songs in anyone's possession anywhere in Varmia. On April 23, Giese dispatched a copy of the *First Account* to Duke Albrecht at his palace in Königsberg. Writing in German to introduce the little Latin treatise, Giese asked "that Your Princely Eminence might look graciously upon this highly learned guest on account of his great knowledge and skill, and grant him your gracious protection."

Rheticus apparently came under the duke's aegis soon afterward, because he remained in residence. Except for a brief return visit to Wittenberg to give two lectures late in 1540, Rheticus continued working alongside Copernicus. Together they reorganized and rewrote several sections of *On the Revolutions*. They reviewed all the demonstrations describing the planetary motions and the directions for deriving specific positions in celestial latitude and longitude. Rheticus likely assisted Copernicus's measurement of the partial solar eclipse that occurred over Frauenburg on April 7, 1540. Sixteen months later, when another partial solar eclipse visited the region on August 21, 1541—the fourth and last one that Copernicus observed—Rheticus was still at his side.

From conversations sustained over their long intimacy, Rheticus composed the only authorized biography of his teacher. Giese commended this prose portrait of Copernicus, but, unfortunately, Rheticus never published it, and the text disappeared.

However much Rheticus feared for his safety or suffered other anguish during the days and nights he spent in Varmia, he comforted himself with the pleasures of the new astronomy. "This and other like sports of Nature," he reported, "often bring me great solace in the fluctuating vicissitudes of my fortunes, and gently soothe my troubled mind."

In August 1540, several months after the *First Account* appeared, the expert printer Petreius penned an open letter to Rheticus and published it as an appendix to an astrology text he issued. Petreius hailed Rheticus for having traveled "to the farthest corner of Europe" to find Copernicus, and for writing such a "splendid description" of his system. "Although he does not follow the common system by which these arts are taught in the schools, nevertheless I consider it a glorious treasure if some day through your urging his observations will be imparted to us, as we hope will come to pass." This encouragement was tantamount to an imprimatur from Petreius. A preeminent press—the leading printer in Nuremberg—stood ready to publish *On the Revolutions*.

Copernicus, however, had not yet committed himself to publication, but only to teaching Rheticus the intricacies of his theory. He devoted considerable time to instructing and sheltering his new disciple, while still shouldering his various administrative duties for the chapter. In September 1540, he registered with Rome his official request for a coadjutor. He was sixty-seven years old, and wished to see his young Danzig relative Jan Loitz, a boy of twelve, groomed to assume his canonry.

As part of the campaign to win Albrecht's protection of Rheticus, Copernicus had offered to make his medical acumen available to the duke on demand. Albrecht found occasion to test this promise in April of 1541, when he wrote to say that "Almighty Sempiternal God is inflicting on one of my counselors and subordinates an affliction and severe illness which does not get better." That same day, April 6, Albrecht also alerted the Varmia Chapter to the situation, in the expectation they would excuse Copernicus to make a house call. The chapter consented on April 8, expressing all the canons' sympathy and announcing that Copernicus, "without any troublesome excuse, at his advanced old age," would gladly comply.

Copernicus left immediately for Königsberg to do Albrecht's bidding. Rheticus went along, as he could hardly remain in Frauenburg without his teacher—or pass up the opportunity to meet his royal patron. Soon after Duke Albrecht received them, he informed the chapter that Copernicus would need to stay a long while by the sick counselor's bedside, "recalling that it is quite Christian and praiseworthy in such a case to act as a fellow-sufferer." During the three weeks Copernicus tended to the invalid, he conferred by mail with the king's physician in Krakow. Neither doctor could do much to soothe the patient, but at least he lived, and Albrecht felt thankful for that. Meanwhile Rheticus and Albrecht explored their mutual interests, which included mathematics, maps, and mapmaking.

Letters from Germany awaited both Rheticus and Copernicus when they got back to Frauenburg in May. Andreas Osiander had written to each of them individually, answering their requests for his advice. Osiander's standing as both a theologian and an amateur mathematician—and a friend of the printer Petreius—uniquely qualified him to consult on how to publish Copernicus's book without offending religious or Aristotelian sensibilities. To Copernicus, he suggested writing an introduction to make the point that mathematical hypotheses "are not articles of faith but the basis of computation; so that even if they are false it does not matter, provided that they reproduce exactly the phenomena of the motions."

To Rheticus, he wrote: "The Peripatetics and theologians will be readily placated if they hear that there can be different hypotheses for the same apparent motion; that the present hypotheses are brought forward, not because they are in reality true, but because they regulate the computation of the apparent and combined motion as conveniently as may be; that it is possible for someone else to devise different hypotheses; that one man may conceive a suitable system, and another a more suitable, while both systems produce the same phenomena of motion; that each and every man is at liberty to devise more convenient hypotheses; and that if he succeeds, he is to be congratulated. In this way they will be diverted from stern defense and attracted by the charm of inquiry; first their antagonism will disappear, then they will seek the truth in vain by their own devices, and go over to the opinion of the author."

A reprint of the popular *First Account* came out in Basel in 1541. This version prominently displayed Rheticus's name on the title page. It also included an introduction by an old family friend of his, the physician Achilles Pirmin Gasser, predicting that the "contrary" sounding content would eventually establish "a true system of astronomy." Although the reprint reached out to a wider audience of mathematicians, its real target seems to have been Copernicus himself—to erode the last shred of his reluctance to publish *On the Revolutions*. Gasser prophesied that support for the *First Account* would prompt "a greater stream of requests" to reach the author of "that rare and almost divine work (whose contents are here adumbrated)," thus "imploring him to allow the delivery of his whole work to us by means of the persistence, effort, and tireless diligence of my friend."

The stream of requests indeed flowed swiftly. Even Bishop Dantiscus received an importuning letter from abroad—from Gemma Frisius, the polymath and instrument maker he had met in the Low Countries during his days as a diplomat. "Urania seems to have a new residence there with you, and raised up new worshippers who are about to offer us a new Earth, a new Sun, new stars, indeed a whole new world," Gemma wrote. "I am filled with desire to see this business brought to fruition. And everywhere there are more than a few erudite men whose minds desire it no less than I do." Dantiscus, inured by now to the Rheticus-Copernicus collaboration, dropped his air of grudging forbearance and finally got behind their project in earnest. In June of 1541, after meeting personally with Copernicus in Braunsberg, Dantiscus composed verses to be used as a foreword for the work in progress.

"I have received your Most Reverend Lordship's very gracious and quite friendly letter," Copernicus acknowledged. "Together with it you did not disdain to transmit also a truly elegant and relevant epigram for the reader of my book." Copernicus promised to place the poem "in the forefront of my work, provided that the work is worthy to deserve being so highly embellished by your Most Reverend Lordship. Yet people who know more than I do, and to whom I should listen, say over and over again that my work is not negligible." Even as he conferred with Rheticus throughout the summer to further revise and expand the text, he still felt qualms about publishing.

The bishop's rhapsody undoubtedly loses much in translation, but says in part:

These writings show you the way to the heavens,
If you want to grasp with your mind the boundaries
Where the very beautiful universe expands its immense spaces,
Or the region of the heavens where the planets wander,
And the changes which their perpetual courses undergo . . .

Rheticus, a sometime poet himself and member of a poets' circle in Wittenberg, resisted comment. With the bishop at last in his corner, he continued to court the duke. In August 1541 he sent Albrecht a copy of a booklet he had written about mapmaking, in German, called *Chorography*. The next day he sent another gift—most likely a gnomon, or shadow-casting instrument for gauging the length of days—along with a letter that begged a favor. The time had come for him to head home to Wittenberg, and he was not at all sure what sort of welcome awaited him there. A word from the duke would guarantee his reinstatement on the faculty, and also buy him time to see Copernicus's book through production. Albrecht obligingly dictated a letter on September 1 to be sent to the Elector of Saxony, John the Magnanimous, with a copy to the university administration.

"Highborn prince, dear affectionate uncle and brother-in-law," Albrecht called John. "Our especially beloved Georg Joachim Rheticus, professor of mathematics in Wittenberg, spent some time reputably and well here in these lands of Prussia. He also pursued his science of astronomy etc. in the same manner with divine grace and help. . . . Accordingly, it is our friendly request to Your Highness that in recognition of his skill, ability, and value, you may wish to validate and confirm him in the aforementioned professorship which he formerly held in Wittenberg. You may also wish graciously to allow and permit him to betake himself for a time, without interruption of his professor's wages, for the sake of carrying out his intended work in the place where he decided to have his book printed. For our sake also, you may wish to show and evince to him all the gracious, beneficial good will, of which we have no doubt."

Thus armed, Rheticus packed up a fair copy of Copernicus's manuscript and said farewell to his teacher. Each knew he would never see the other again, and their emotions at parting probably blended grief with a measure of relief.

"Upon my departure," Rheticus later remembered, "the great old man solemnly charged me to carry on and finish what he, prevented by old age and impending death, was unable to complete by himself." In another context he wrote, "There has been no greater human happiness than my relationship with so excellent a man and scholar as he."

When Rheticus arrived back in Wittenberg in October, the university immediately made him dean of the Faculty of Arts. This was probably as thankless a job then as now, and it burdened him with responsibilities that impeded his publishing agenda. He also acquired a new epithet, "Joachim Heliopolitanus," or "one who comes from the City of the Sun." Less a compliment than a mild condemnation of his adopted cosmology, the title clicked with an anti-Copernican comment Martin Luther purportedly dropped at lunch one day.

"So it goes now," the great Reformer was overheard to remark. "Whoever wants to be clever must agree with nothing that others esteem. He must do something of his own. This is what that fellow does who wishes to turn the whole of astronomy upside down. Even in these things that are thrown into disorder I believe the Holy Scriptures, for Joshua commanded the Sun to stand still and not the Earth." Another diner recalled Luther's branding Copernicus a "fool," and perhaps he did use that word, though all the "Table Talk" is hearsay. Melanchthon, on the other hand, wrote a letter in October 1541 that expressed his annoyance with "the Polish astronomer who moves the Earth and immobilizes the Sun."

Between teaching duties and presiding at graduation ceremonies in February and April, Rheticus could not break away to take Copernicus's manuscript to Nuremberg. Perhaps out of frustration, he selected two chapters treating technical aspects of geometry, called them *On the Sides and Angles of Triangles*, and published them in Wittenberg in 1542 with full credit to "the most illustrious and highly learned Mr. Nicolaus Copernicus." At the opening of this book, he included—or rather, unloaded—the poem by Dantiscus, without attribution.

Not till early May 1542, after his term as dean ended, did Rheticus arrive in Nuremberg to deliver the bulk of the copied manuscript to Petreius. Printing commenced immediately. By the end of the month Rheticus had already corrected the first two signatures of eight pages each. In August, with progress well under way, he reflected on his Frauenburg adventure: "I regret neither the expense, nor the long journey, nor any of the other hardship. Rather, I feel I have reaped a great reward, namely that I, a rather daring youth, compelled this venerable man to share his ideas sooner in this discipline with the whole world. And all learned minds will join in my assessment of these theories as soon as the books we now have in press in Nuremberg are published."

Chapter 8

On the Revolutions of the Heavenly Spheres

I confess that I shall expound many things differently from my predecessors, although I shall do so thanks to them, and with their aid, for it was they who first opened the road of inquiry into these very questions.

—From Copernicus's introduction to Book I, *On the Revolutions of the Heavenly Spheres,* 1543

LONE AGAIN WITH HIS FEARS of ridicule after Rheticus left, Copernicus fussed over his original manuscript. He jotted notes in the margins regarding a few new thoughts and corrections. He felt grave misgivings about the Mercury sections in Books V and VI. Even the observations Rheticus had brought from Schöner proved largely useless in constraining Mercury's orbit to his system, and Copernicus wound up with a clumsy adaptation of Ptolemy's model for the innermost planet.

In mid-June 1542 Pope Paul III approved the choice of young Jan Loitz as Copernicus's coadjutor. The news, in the form of a papal writ, did not reach Varmia for many months, however, and in the interim Copernicus drafted a long letter to His Holiness on a different matter. Although he clearly addressed this letter to Pope Paul at the Vatican, he sent the final draft to Rheticus, care of Petreius in Nuremberg, to serve as the dedication for *On the Revolutions*.

"I can readily imagine, Holy Father," he began, "that as soon as some people hear how in this volume, which I have written about the revolutions of the spheres of the universe, I ascribe certain motions to the terrestrial globe, they will immediately shout to have me and my opinion hooted off the stage." His reluctance to publish had never fully subsided. Even now, he avowed, he had agreed to publish his work only at the repeated urging of insistent friends.

"They exhorted me no longer to refuse, on account of the fear which I felt, to make my work available for the general use of students of astronomy. The crazier my doctrine of the Earth's motion now appeared to most people, their argument ran, so much the more admiration and thanks would it gain after the publication of my writings dispelled the fog of absurdity by most luminous proofs."

Copernicus nowhere recorded the circumstances of his decision to dedicate his book to Pope Paul. No notes from Giese, Dantiscus, or any other dignitaries so much as hint as to how the idea came about, or how they obtained permission from the papal curia. His Holiness Paul III, né Alessandro Farnese, possessed no personal knowledge of mathematics himself, but expressed his interest in the uses of that science through the employment of a full-time, high-profile astrologer, Luca Gaurico. In 1534, as thanks to Gaurico for predicting Paul's ascent to the throne of St. Peter, the new pope invited the favored astrologer to Rome and made him a bishop.

Copernicus credited Paul with at least a partial understanding of the motions of the heavenly spheres. In the dedication letter, he walked the Holy Father quickly through the unsatisfying homocentrics, eccentrics, and epicycles that had failed to produce "the structure of the universe and the true symmetry of its parts."

"After long reflection," he continued, "I began to be annoyed that the movements of the world machine, created for our sake by the best and most systematic Artisan of all, were not understood with greater certainty by the philosophers." Copernicus had circumvented the schools of astronomy, he said, to reread all of philosophy. In the pages of Cicero and Plutarch, he had found references to those few thinkers who dared to move the Earth "against the tradi-

Alessandro Farnese, elected Pope Paul III in 1534, in a painting by Titian.

tional opinion of astronomers and almost against common sense." (He still knew nothing of the Earth-moving plan of Aristarchus, which had not yet been reported to Latin audiences.)

"Therefore I too began to consider the mobility of the Earth. And even though the idea seemed absurd, nevertheless I knew that others before me had been granted the freedom to imagine."

Thus liberated, he had correlated all the heavenly motions, as confirmed in the present volume, the contents of which he summarized before laying the work at the pope's feet.

"In order that the educated and uneducated alike may see that I do not run away from judgment, I have preferred dedicating these results of my nocturnal study to Your Holiness rather than to anyone else. For even in this very remote corner of the Earth where I live, you are considered the highest authority by virtue of the loftiness of your office and your love for all literature and even of mathematics."

With that as preamble, he came to the real need for papal protection: "Perhaps there will be babblers who claim to be judges of astronomy although completely ignorant of the subject and, badly distorting some passage of Scripture to their purpose, will dare to find fault with my undertaking and censure it." He and Rheticus had often discussed this possibility with Giese. They anticipated how Joshua's command for the Sun to stand still might be hurled at Copernicus to prove the Sun's motion and thereby destroy his whole theoretical edifice. Or criticism could wrap itself in Psalm 93's proclamation that the foundations of the Earth remain forever unmoved—or Ecclesiastes' account of how the Sun moves from sunrise to sunset and then hastens back to its rising place. Against the likelihood of such a biblical backlash, Rheticus had prepared a tract in which he rectified Holy Scripture with the Copernican ideal, but he had not yet published it. Even if Rheticus's defense did appear in print, it could never approach the power of a single word from the pope.

"Astronomy is written for astronomers," Copernicus asserted at the end of the dedication letter, for the simple reason that they alone could follow the mathematical proofs. That self-same audience of astronomers would recall the efforts of Leo X and the Lateran Council to reform the ecclesiastical calendar. They would remember how the attempt had failed for lack of adequate measurement of "the lengths of the year and month and the motions of the Sun and Moon." Ever since that time, Copernicus said, "I have directed my attention to a more precise study of these topics. But what I have accomplished in this regard, I leave to the judgment of Your Holiness in particular and of all other learned astronomers. And lest I appear to promise more about the usefulness of this volume than I can fulfill, I now turn to the work itself."

At the time Copernicus concluded this dedication, in June 1542, the first few sections of typeset text—chapters 1 through 6 of Book I—arrived in Frauenburg for his inspection. They looked quite good. Petreius had chosen an attractive roman font, with large, elegantly decorated initial capital letters by the distinguished Nuremberg artist Hans Sebald Beham to inaugurate each chapter. The lone geometri-

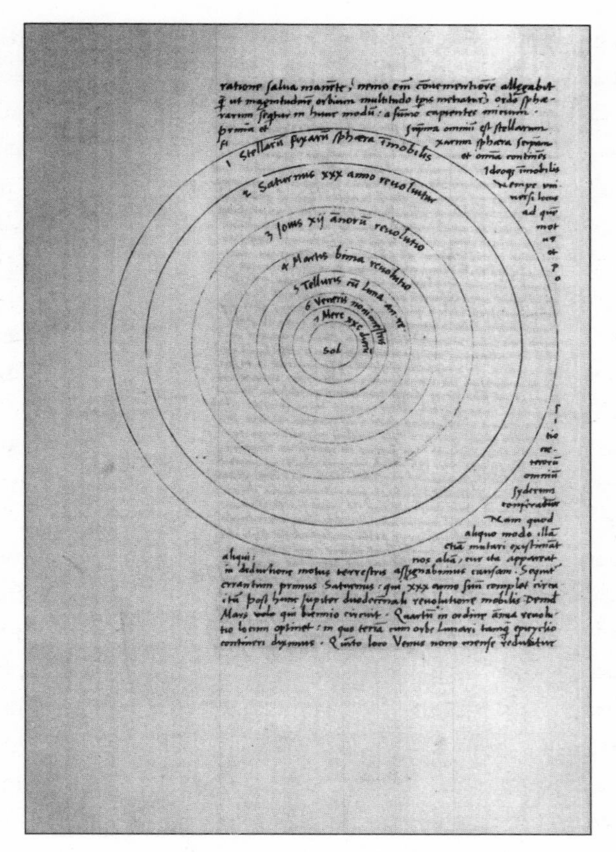

While drawing the several spheres of the planets with a pair of compasses, Copernicus inadvertently drilled a small hole in this page of his manuscript copy of On the Revolutions.

cal figure in these introductory passages appeared crisp and clear—evidence that the printer had commissioned skilled carvers to cut woodblocks for the 142 required diagrams. Petreius had volunteered to absorb this and all other production costs, including more than a hundred reams of *P*-watermarked paper to run off the several hundred intended copies of the book. Still, Copernicus could not help finding a few infelicities he wished to correct, and which he pointed out by return mail. Although Petreius could not halt progress to reprint every page that Copernicus amended, he noted many of the author's changes on an errata leaf printed later.

Rheticus hovered over the press, proofreading. This might not have been a truly full-time job, since he read so much faster than the type could be set and inked in the large flat plate, the paper positioned and impressed, and the double-spread printed sheets hung up to dry on both sides. The slow pace of perhaps two pages per day, further retarded by the wait for woodcuts or other delays, allowed Rheticus a few weeks off that summer. Two visits to family and friends in and around Feldkirch, once in early June and again in September, barely diverted him from his duty to *On the Revolutions*. Nor did he shirk his primary responsibility when he prepared two of his own recent commencement orations for publication by Petreius in August.

However, the time Rheticus devoted to negotiating a new teaching position—and his success on this score—abruptly ended his career as Copernicus's proofreader. In mid-October, with less than half the book done, he left Nuremberg to accept the professorship of higher mathematics at the University of Leipzig, two hundred miles away from the print shop. Whereas before Rheticus had taught lower, or rudimentary, mathematics at Wittenberg, he would now lecture on advanced astronomy. He also realized a large financial gain over his former salary. Leipzig University's records for 1542 state that when Rheticus refused the standard professorial pay of 100 florins per year, the authorities increased the offer to 140 in order to entice him.

Although no correspondence between Rheticus and Copernicus survives from this (or any other) juncture, it seems likely the disciple would have informed his teacher that he had moved on and turned over the proofreading responsibility to someone else—to Andreas Osiander, in fact.

Osiander had a history with Rheticus and Copernicus. His letters of the year before bespoke his great interest in their publishing venture, though his religious beliefs colored his opinion of all astronomical models and hypotheses. As he had told them, any number of competing sets of conjectures might account for the observed heavenly motions, but nothing short of divine revelation could determine which set, if any, truly corresponded to reality. And so, since it was impossible to know the truth, one astronomer should refrain from insulting another by insisting he had uncovered the actual workings of the celestial spheres.

Osiander also had ties to *On the Revolutions* through Petreius, who was an acquaintance of several years' standing. Petreius had published some of Osiander's sermons, and occasionally called on his services as an editor as well as a proofreader. It is unclear whether Rheticus or Petreius chose Osiander to fill the vacant slot, though they might well have shared the same good opinion of his qualifications.

Copernicus continued to receive batches of pages from the press all this while, although, after November 1542, he could no longer read and comment on them. In late autumn, at age sixty-nine, he suffered a stroke—a cerebral hemorrhage that raided his memory,

Andreas Osiander, Minister of St. Lorenz Church in Nuremberg.

robbed him of speech, and paralyzed the right side of his body. His friend Jerzy Donner, who had joined the chapter as a canon two years previously, alerted Giese.

"I was shocked by what you wrote about the impaired health of the venerable old man, our Copernicus," Giese replied on December 8, 1542. "Just as he loved privacy while his constitution was sound, so, I think, now that he is sick, there are few friends who are affected by his condition. I therefore ask you . . . please to watch over him and take care of the man whom you cherished at all times together with me. Let him not be deprived of brotherly help in this emergency."

At the end of December, when the news of Copernicus's infirmity reached his relatives in Danzig, Jan Loitz's father reminded Bishop Dantiscus that the boy stood ready to take possession of the fourteenth Varmia canonry—Copernicus's canonry—as soon as Rome approved.

Canon Fabian Emerich, the chapter's replacement physician, judged the medical situation hopeless. Copernicus could do little but lie in bed, and he ate hardly anything. Attended by Donner through the winter and spring, Copernicus gradually declined, drifting in and out of consciousness until early May, when he no longer woke and slept but stayed asleep continuously. On May 24, 1543, the final pages of his book arrived from Nuremberg. Donner took them to the invalid's bedside, put them in his hands, and the next moment saw the life go out of him—as though Copernicus had held on all those months just to see the thing complete, and now he could let go.

They buried him, as was customary, in the sandy soil under the floor of the cathedral, somewhere near his own altar. No marker or epitaph designated his exact resting place, but that, too, was the custom.

His will divided his cash reserve of five hundred marks among the children of his nieces—Katyryna's daughters, long since married and mothers several times over. If he accumulated greater wealth through the years, he must have given it away before he fell ill. He left his medical texts to Emerich, and his other books to the chapter library. His own book, his only lasting legacy, was now an orphan.

The final signatures of *On the Revolutions* to reach Copernicus contained the first few pages, including the title page, which identified the author without fanfare as "Nicolaus Copernicus of Torun." After a lifetime spent in Varmia, he still belonged to his native city, while his work, *Six Books on the Revolutions of the Heavenly Spheres*, entered the world (as the bottom of the page attests) through the press of Johannes Petreius, Nuremberg, 1543. Above his own name the printer had placed a welcome—and a warning—to the potential audience.

"You have in this recent work, studious reader," Petreius announced, "the motion of both the fixed stars and the planets, restored on the basis of ancient as well as recent observations, and also outfitted with new and marvelous hypotheses. You also have most expeditious tables, from which you can compute those motions with the utmost ease for any time whatever. Therefore buy, read, profit."

The very next line sounded a caution in Greek: "Let no one untrained in geometry enter here." This maxim, which reputedly appeared over the gate to Plato's Academy, reiterated Copernicus's own contention that mathematics was written for mathematicians.

Turning over the title page, brave readers encountered yet another caveat, under the heading "To the Reader Concerning the Hypotheses of this Work." This note anonymously acknowledged the clamor surrounding the publication of the book in hand:

"There have already been widespread reports about the novel hypotheses of this work, which declares that the Earth moves whereas the Sun is at rest in the center of the universe. Hence certain scholars, I have no doubt, are deeply offended and believe that the liberal arts, which were established long ago on a sound basis, should not be thrown into confusion. But if these men are willing to examine the matter closely, they will find that the author of this work has done nothing blameworthy. For it is the duty of an astronomer to compose the history of the celestial motions through careful and expert study. Then he must conceive and devise the causes of these motions or hypotheses about them. Since he cannot in any way attain to the true causes, he will adopt whatever suppositions enable the motions to be computed correctly from the principles of geometry for the future as well as for the past. The present author has performed both these duties excellently. For these hypotheses need not be true nor even probable. On the contrary, if they provide a calculus consistent with the observations, that alone is enough."

There followed a familiar trope: "And if any causes are devised by the imagination, as indeed very many are, they are not put forward to convince anyone that they are true, but merely to provide a reliable basis for computation. However, since different hypotheses are sometimes offered for one and the same motion . . . the astronomer will take as his first choice that hypothesis which is the easiest to grasp. The philosopher will perhaps rather seek the semblance of the truth. But neither of them will understand or state anything certain, unless it has been divinely revealed to him."

Some readers assumed these words to be Copernicus's own. Others recognized them as the voice of another, but were left to guess at its identity as they continued reading.

"Therefore alongside the ancient hypotheses, which are no more probable, let us permit these new hypotheses also to become known, especially since they are admirable as well as simple and bring with them a huge treasure of very skillful observations. So far as hypotheses are concerned, let no one expect anything certain from astronomy, which cannot furnish it, lest he accept as the truth ideas conceived for another purpose, and depart from this study a greater fool than when he entered it. Farewell."

Although Copernicus himself had at last loosed his vision of "the composition of movements of the spheres of the world," this anonymous preamble reduced his effort to the status of an interesting and worthy aid to calculation, wholly unrelated to reality.

Chapter 9
The Basel Edition

Anyone can rightly wonder how, from such absurd hypotheses of Copernicus, which conflict with universal agreement and reason, such an accurate calculation can be produced.

—Anonymous handwritten note in an early copy of *On the Revolutions*

When Rheticus received his teacher's finished book—when he realized that Osiander had won his way with it after all—he threatened to "so maul the fellow that he would mind his own business and not dare to mutilate astronomers any more in the future." But he could neither prove Osiander's complicity nor deny his own. Had he stayed at the print shop, he might have prevented this outcome. And so, with anger directed perhaps as much inward as outward, Rheticus defaced several copies of the book that came into his hands. First he crossed out part of the title with a red crayon, suggesting that "of the Heavenly Spheres" had been wrongly tacked on as an unauthorized addendum to the intended *On the Revolutions*—possibly to divert attention from the motion of the Earth. Then Rheticus put a big red *X* through the entirety of the anonymous note "To the Reader." The crayon cross did not hide the demeaning message, however. Giese could still read it plainly in both copies of the book that Rheticus sent him, which he discovered waiting for him at Kulm, draped in the news of Copernicus's death, when he arrived home from the marriage celebrations of Crown Prince Sigismund Augustus and Archduchess Elisabeth of Austria.

"On my return from the royal wedding in Krakow I found the two copies, which you had sent, of the recently printed treatise of our Copernicus. I had not heard about his death before I reached Prussia. I could have balanced out my grief at the loss of that very great man, our brother, by reading his book, which seemed to bring him back to life for me. However, at the very threshold I perceived the bad faith and, as you correctly label it, the treachery of that printer, and my anger all but supplanted my previous sorrow."

Giese could not decide whether to blame Petreius or someone who worked under him—some "jealous person" who feared that Copernicus's book would achieve the fame it deserved, thereby forcing mathematicians to abandon their previous theories. Still, Giese insisted that Petreius bear the guilt and be punished for his crime.

"I have written to the Nuremberg Senate, indicating what I thought must be done to restore faith in the author. I am sending you the letter together with a copy of it, to enable you to decide how the affair should be managed. For I see nobody better equipped or more eager than you to take up this matter with the Senate. It was you who played the leading part in the enactment of the drama, so that now the author's interest seems to be no greater than yours in the restoration of this work, which has been distorted."

Giese urged Rheticus to demand the opening pages be printed anew, and to include a new introduction by Rheticus, to "cleanse the stain of chicanery."

"I should like in the front matter also the biography of the author tastefully written by you, which I once read," Giese said. "I believe that your narrative lacks nothing but his death on May 24. This was caused by a hemorrhage and subsequent paralysis of the right side, his memory and mental alertness having been lost long before. He saw his treatise only at his last breath on his dying day."

Giese suggested that Rheticus also incorporate in the new

NICOLAI CO-
PERNICI TORINENSIS
DE REVOLVTIONIBVS ORBI-
um coelestium, Libri VI.

Habes in hoc opere iam recens nato, & ædito, studiose lector, Motus stellarum, tam fixarum, quàm erraticarum, cum ex ueteribus, tum etiam ex recentibus obseruationibus restitutos: & nouis insuper ac admirabilibus hypothesibus ornatos. Habes etiam Tabulas expeditissimas, ex quibus eosdem ad quoduis tempus quàm facillime calculare poteris. Igitur eme, lege, fruere.

Ἀγεωμέτρητος οὐδεὶς εἰσίτω.

Norimbergæ apud Ioh. Petreium,
Anno M. D. XLIII.

Rheticus inscribed this copy of On the Revolutions *to Jerzy Donner, the Varmian canon who cared for Copernicus in his final days.*

introduction "your little tract, in which you entirely correctly defended the Earth's motion from being in conflict with the Holy Scriptures. In this way you will fill the volume out to a proper size and you will also repair the injury that your teacher failed to mention you in his Preface."

Copernicus's preface, addressed as it was to Pope Paul III, could hardly have acknowledged his Lutheran assistant. But Giese had just seen himself named in the preface as the friend who overcame Copernicus's reluctance to publish, and he must have felt sheepish receiving the lion's share of credit for what had truly been Rheticus's doing. "I am not unaware," he reminded Rheticus, "how much he used to value your activity and eagerness in helping him. . . . It is no secret how much we all owe you for this zeal." It was no secret, and yet Rheticus remained anonymous while the preface touted Giese as "a man who loves me dearly, a close student of sacred letters as well as of all good literature," who "repeatedly encouraged me and, sometimes adding reproaches, urgently requested me to publish this volume and finally permit it to appear." The preface thanked only one other person by name—the now deceased Cardinal of Capua, Nicholas Schönberg, whose laudatory, signed letter of 1536 had been exhumed from Copernicus's files and printed in full as part of the front matter. Alas, Rheticus, who had contributed the most, was lumped of necessity with "not a few other very eminent scholars" whom Copernicus acknowledged in a single nod. Giese stumbled over himself apologizing to Rheticus: "I explain this oversight not by his disrespect for you, but by a certain apathy and indifference (he was inattentive to everything which was nonscientific) especially when he began to grow weak."

In closing, Giese asked whether Rheticus or anyone else had sent the book to the pope, "for if this was not done, I would like to carry out this obligation for the deceased."

Rheticus followed all of Giese's instructions. As a result, the Nuremberg Senate issued a formal complaint against Petreius, but the

Johannes Petreius, citizen and printer of Nuremberg.

printer pled innocence. He insisted that the front matter had been given to him exactly as it appeared, and he had not tampered with it. Petreius used such heated language in his statement of self-defense that the Senate secretary suggested his "acerbities" be "omitted and sweetened" before forwarding his comments to the Bishop of Kulm. The Senate, taking Petreius at his word, decided not to prosecute him. And no revised edition of *On the Revolutions* ever emerged from his press.

Several times that summer of 1543, while Giese and Rheticus sought to defend their friend's honor, Anna Schilling returned to Varmia. Although she had moved to Danzig after Bishop Dantiscus banished her from the diocese, she still owned a house in Frauenburg. Perhaps, now that Copernicus had passed on, she expected no objection to her presence. Her visits each lasted a few days, allowing her time to remove her belongings and find a buyer. On September 9, she at last sold the property. The next day the officers of the chapter, who had been monitoring her movements all along, reported her to the bishop. They wanted to know whether she should continue to suffer exclusion from

Varmia, given that the legal cause of her banishment had vanished at Copernicus's death. It would seem she planned to leave and never return, having liquidated her last ties to the region, but still the canons posed the question, and Dantiscus rapidly replied.

"She, who has been banned from our domain, has betaken herself to you, my brothers. I am not much in favor, whatever the reasons. For it must be feared that by the methods she used to derange him, who departed from the living a short while ago, she may take hold of another one of you. . . . I would consider it better to keep at a great distance, rather than to let in, the contagion of such a disease. How much she has harmed our church is not unknown to you."

From Leipzig, Rheticus sent personally inscribed, red-crayoned copies of *On the Revolutions* as gifts to his friends at Wittenberg. Their reaction to Copernicus differed markedly from the disciple's own. Melanchthon, as intellectual leader of the faculty, followed Luther's lead by spurning the new order of the planets on biblical grounds. One wonders whether Melanchthon ever read "the little tract" by Rheticus that Giese liked so much—the one in which Rheticus "entirely correctly defended the Earth's motion from being in conflict with the Holy Scriptures." If he did read it, he was not at all swayed by its arguments. At the same time, however, Melanchthon recognized the value of Copernicus's contribution to planetary position finding, and commended Copernicus's improved analysis of the Moon's motion. The Wittenberg mathematicians—Rheticus's former colleagues, Erasmus Reinhold and Caspar Peucer—echoed Melanchthon's response. They skimmed over the heliocentric universe laid out in Book I of *On the Revolutions*, reserving their careful scrutiny for the technical sections that came afterward. They rejoiced in the way Copernicus righted Ptolemy's wrongs by returning uniform circular motion to the heavenly bodies, but they rejected the Earth's rotation and revolution. They ignored the reordering of the spheres, along with all the new idea's implications for the distances to the planets and the overall size of the universe.

Reinhold immediately began constructing new tables of planetary data, based entirely on Copernicus's devices. Although Copernicus had provided various tables in *On the Revolutions*, many of the numbers needed for calculating planetary positions lay scattered throughout the text. Reinhold gathered all this information into convenient form for working astronomers—that is, for astrologers. Melanchthon blessed Reinhold's effort, then enlisted financial backing for its publication from Duke Albrecht, who obliged. It seemed only fitting that Reinhold name his project the *Prutenic Tables*, in honor of Prussia, home to both Copernicus and Albrecht.

No one can say what effect Rheticus might have exerted on the Wittenberg scholars had he remained among them, but it seems unlikely he could have defended Copernicus's cosmos against the criticisms of Luther and Melanchthon. Absent Rheticus, Copernicus's intent, which had already suffered the undermining of Osiander's note to readers, underwent further subversions at Wittenberg. Reinhold's published *Tables* meshed the planetary models with a stationary, central Earth. Peucer, in his book *Hypotheses astronomicae,* reinstated the ninth and tenth spheres beyond the fixed stars.

After two years at Leipzig, Rheticus, restless again, left his students without permission in the autumn of 1545 to visit the mathematician and astrologer Girolamo Cardano in Milan. Rheticus took along a gift copy of Copernicus's book, which he inscribed to Cardano when he arrived.

The two had previously collaborated by correspondence on a collection of genitures (horoscopes) of famous men, published by Petreius in the same year as *On the Revolutions*. Now Cardano was preparing an enlarged new edition, and Rheticus gave him several detailed horoscopes for inclusion. One concerned Andreas Vesalius, the physician whose 1543 masterpiece, *On the Fabric of the Human Body*, had corrected ancient misconceptions, thereby doing for anatomy what *On the Revolutions* did for astronomy. Another geniture sketched the character and life circumstances of the mathematical prodigy Johannes Regiomontanus, who wrote the *Epitome of Ptolemy's Almagest*—the book Copernicus had studied so closely in his youth.

Rheticus also delivered Osiander's horoscope to Cardano at this time. He had come with the hope that Cardano might help him in return, perhaps as a collaborator on the great project Rheticus envisioned about the science of triangles. Instead, Rheticus found Cardano trapped in the triangular compartments of his horoscope diagrams, extracting the most detailed predictions from them, down to the manner of a man's death and the disposition of his corpse. While these techniques surely fascinated Rheticus, his Milan sojourn soured under Cardano's personal slights and lukewarm response to *On the Revolutions*.

That summer, Leipzig summoned Rheticus back to his teaching duties after the year's unauthorized absence. He left Italy for Leipzig in autumn 1546 but stopped en route in Lindau, where he suffered a mysterious nervous and physical breakdown that deranged him for several months. Fortunately his former Wittenberg classmate Caspar Brusch, the Lindau schoolmaster, looked after Rheticus and later recounted his ordeal in a letter to a mutual friend of theirs:

"He has halfway regained his former state of health (after having been, while here with me, severely ill), though he is not yet fully recovered. . . . I know that something about an evil spirit has been rumored abroad by commercial travelers. Even if this report is not quite false or devoid of substance, still Joachim is taking it very hard, since he is apprehensive that it could somewhat damage his former reputation." Rheticus's mother, a Catholic, also lived in the area, in Bregenz, with her wealthy second husband. She urged her son to seek release from his demons by making a pilgrimage to St. Eustatius's shrine in Alsace, but he refused.

"He lay ill here for nearly five months," Brusch continued in his letter of late August 1547, "and every day I would visit, conversing and fellowshipping with him. Throughout this time I made available to him the whole Bible, in both German and Latin, from my library, as well as many of the devotional writings of Luther, Melanchthon, and Cruciger. These he read and reread so diligently that in the end he knew them through and through. However, on particular occasions, with a full heart and most fervent vows, indeed often in tears, he would call upon the Son of God, awaiting deliverance from him alone."

When his crisis passed, Rheticus further postponed his return to Leipzig by spending the winter in Zurich, still striving to recover his health. While there, he wrote a proposal for improving the triquetrum, now that the observational instrument was enjoying a new vogue in the wake of *On the Revolutions*. Copernicus had apparently "stimulated certain eminent men to observe the motions of the heavenly bodies," Rheticus said. He published his triquetrum suggestion in Zurich, in February 1548, but dedicated it "to the teachers and professors of the faculty of arts at the University of Leipzig." That same month, on February 13, he wrote to them directly, saying he would see them before long. On the sixteenth, he turned thirty-four—surprised, perhaps, to find himself still alive. He spent Easter in Baden, bathing in the thermal springs on his doctor's advice, and finally reentered Leipzig at summer's end. As happened when he returned to Wittenberg following his first protracted leave, he found the deanship of the Faculty of Arts thrust upon him.

Despite the duties of administration and teaching, Rheticus rebounded during this period. In October 1549, he wrote to Giese to report progress on several new works relevant to astronomy. He hoped his soon-to-be-published calendar of prognostications for the coming year would sell well enough to fund his private researches on more serious topics. For example, he had just completed a modern scientific commentary on the geometry classic, Euclid's *Elements*, and would soon issue his *Ephemerides: A Setting Forth of the Daily Position of the Stars . . . by Georg Joachim Rheticus according to the theory . . . of his teacher Nicolaus Copernicus of Torun*. He never tired of reminding readers that it was Copernicus "whose hand advanced the machinery of this world." As heir to that tradition, "I have not wanted to backslide from Copernican teaching, not even by a finger's width."

Through his resumed communication with Giese, Rheticus discovered Dantiscus had died and ceded the Varmia bishop's seat to his old rival. But Giese lasted only a year in that pinnacle position, before his own death in October 1550, at age seventy. Then Stanislaw Hozjusz, Dantiscus's chosen, took over as Bishop of Varmia, and fought so strenuously against the Lutheran heresy—even winning some prominent converts back to Catholicism—that he was elevated to cardinal.

At the start of 1551, Rheticus chose the word *canon*—reminiscent of Copernicus, but also signifying a code of instructions—for the title of his pamphlet *Canon of the Science of Triangles*. It offered the best available trigonometric tables—crucial calculation aids for astronomers. Lest anyone miss this slim volume's relation to Copernicus, Rheticus cemented the connection in a playful introductory dialogue between "Philomathes," a lover of math, and his guest "Hospes." When Hospes asks, "What sort of man is this Rheticus?" Philomathes replies, "He indeed is the one who is now delivering to us this fruit from the most delightful gardens of Copernicus. For after his recent return from Italy he resolved to impart freely to students of mathematics everything he learned from that excellent old man, as well as everything he has acquired by means of his own effort, perseverance, and devotion."

The question "What sort of man is this Rheticus?" came up again in the spring of 1551, when the father of a Leipzig student charged that the professor had perpetrated "a sudden, outrageous, and unchristian" act on his son. After luring the boy, "a minor child" in his father's eyes, Rheticus allegedly "plied him with strong drink, until he was inebriated; and finally did with violence overcome him and practice upon him the shameful and cruel vice of sodomy."

The legal punishment for the crime of sodomy was "death by fire." Guilty or no, Rheticus ran away in April, before the winter term ended or his criminal trial could begin. Rumors placed him in Prague. Numerous urgent letters in fact reached him there, summoning his appearance in court, but Rheticus never returned to Leipzig. After months of fruitless maneuvering, the university sentenced him in absentia to banishment for a period of 101 years. In recognition of his talents, however, and for the sake of decorum, the particulars of his accusation and sentencing were kept confidential.

Rheticus spent that first year in Prague studying medicine at Charles University, and probably the next year as well. Then he continued his medical training in Silesia through 1554. He seems to have seen medicine as the ideal means to guarantee himself an income for as long as he remained unattached to an academic position and without the support of a patron. Then, too, medicine had been his father's profession, and that of Copernicus as well.

"Doctor" Rheticus moved to Krakow in the spring of 1554. He chose the city for its geographical location, three hundred miles due south of Frauenburg, which connected him to Copernicus by a mapmaker's meridian. He resided there for nearly two decades (longer than he ever lived anywhere else), pursuing his old career and his new one simultaneously. On July 20 he wrote to a former student, "I have erected a fifty-foot obelisk in a perfectly level field that the marvelous Mr. Johannes Boner has made available to me for this purpose. By this means, God willing, I shall describe anew the whole sphere of the fixed stars." He shared the work of computing the new, more precise tables he planned to publish with as many as five hired helpers at a time, dividing fractions of angles by a factor of ten billion. Thus engaged in Krakow, Rheticus realized one day that he was past forty, and that he must have erred in forecasting his early demise. The habit of living had gone on for so long, he believed he might reach old age after all. A decade later, in 1563, he "again picked up the work of Copernicus," he said—referring to his teacher's book, as opposed to his legacy—and considered "elucidating it with a commentary."

Heinrich Petri of Basel, a distant relative of Johannes Petreius, published the second edition of *On the Revolutions* in 1566. It contained, in lieu of any new elucidating commentary by Rheticus, the third printing of the disciple's original digest, or *First Account*. If ever Rheticus felt snubbed by lack of acknowledgment in the first edition of *On the Revolutions*, his role in it was here made manifest.

His face stayed hidden, however, and remains so even now. Any formal portrait of him, any sketch or caricature, must have been lost long ago. Despite his numerous publications and acquaintances, his university affiliations, his several occupations, and the curiosity that carried him to so many doors, he left no impression of his physical appearance.

In 1572, Rheticus moved one last time, south from Krakow to the Hungarian city of Cassovia, where a new patron waited to underwrite him. Rheticus persevered with his great opus on the science of triangles, but he was distracted by his medical interests, hampered by having lost important papers through his various relocations, and daunted by the mountain of calculation still to conquer. In a replay of life events he could never have anticipated, he opened his door one spring day in 1574 to a surprise visitor from Wittenberg, who had read Rheticus's pamphlet about triangles (the *Canon*) and heard of his ambition to complete a larger work.

"We had hardly exchanged a few words on this and that," the youth, Valentin Otto, later recalled, "when, on learning the cause of my visit he burst forth with these words: 'You come to see me at the same age as I was myself when I visited Copernicus. If it had not been for my journey, his work would never have seen the light of day.' "

The similar stories diverge at that point. Although Otto proved the most devoted apprentice, he toiled only a short time alongside his teacher. After a few months, Rheticus sent Otto to Krakow to retrieve records he had left there. Returning November 28 from a harrowing journey through rain and flood ("Twice in one day I was in danger of drowning"), Otto found Rheticus critically ill. He tended to his mentor over the next several days, during which time Rheticus officially transferred to him the right—and the duty—to complete the *Science of Triangles*. Otto swore he would, and Rheticus died in his arms early on Saturday, December 4, 1574, at sixty years of age.

Spurred on by the great love he professed for Rheticus, Otto struggled through the next twenty years to complete his inherited endeavor. Almost as soon as the fifteen-hundred-page text appeared, in 1596, its many errors and insufficiencies revealed themselves. Otto had already grown senile, however, and could not rectify the text even when its problems were pointed out to him. A more worthy successor to the work later emerged in the person of Bartholomew Pitiscus, chaplain to the Prince Elector Frederick IV of Heidelberg. After Otto died in 1602, Pitiscus pored through the welter of disorganized notes he had accumulated. "I excavated them one page at a time from their state of neglect, filthy and almost putrid," he said. Through the "irksome" work, he "gleaned many things that have delighted me wonderfully." Pitiscus saw the final version of Rheticus's contribution to the Copernican theory published in Frankfurt in 1613. He titled it *Mathematical Treasury: or, Canon of Sines for a Radius of 1,000,000,000,000,000 Units . . . as Formerly Computed at Incredible Effort and Cost by Georg Joachim Rheticus.*

Chapter 10

Epitome of Copernican Astronomy

I deem it my duty and task to advocate outwardly also, with all the powers of my intellect, the Copernican theory, which I in my innermost have recognized as true, and whose loveliness fills me with unbelievable rapture when I contemplate it.

—Johannes Kepler, *Epitome of Copernican Astronomy*, 1617–21

ONTENT WITH THE *Prutenic Tables*, European astronomers took Copernicus at Osiander's cautious word for the remainder of the sixteenth century—with two monumental exceptions. Between them, the flamboyant Tycho Brahe and the studious, passionately reverent Johannes Kepler carried Copernicus's work to completion.

The Danish Tycho was literally star-struck in 1559, during his thirteenth summer, when a lunar eclipse illuminated the mathematics he was learning at a Lutheran university in Copenhagen. His noble birth gave him the means to purchase his own astronomy books, which he bought secretly, he said, and also read in secret, since his elders considered such pastimes beneath him. Soon he began logging his own observations of the planets and casting the horoscopes of famous men. At twenty-five, after losing most of his nose in a duel, he looked up one November night to see a nova's explosion blaze suddenly to brilliance in the constellation Cassiopeia. He spoke of that 1572 event ever after as the moment when the heavens chose him to be their interlocutor.

"In truth, it was the greatest wonder that has ever shown itself in the whole of nature since the beginning of the world," he announced in his hastily written book, *De nova stella*. Tycho's new star indeed heralded a cataclysm. By its position in the heavens—too far from the zodiac to be a planet, too steadfast for a comet, and supralunar to boot—it boded the end of immutability in Aristotle's celestial realms. Change could occur on high, Tycho's careful observations showed, in the guise of a new star's light. This claim competed for strangeness with Copernicus's moving Earth—and perhaps Tycho winked at Copernicus when he compared the miracle of his *Nova stella* to Joshua's stopping the Sun by prayer.

Tycho took his homeland's far northern latitudes (worse than those bemoaned by Copernicus) as a proud birthright, and dedicated an early work to King Frederick of Denmark. While Tycho did allow that the extreme cold of the climate could disturb an astronomer's serenity, it seems never to have deterred him. Five years after his nova discovery, in the early dark of another November night, Tycho stood fishing at a pond when a comet appeared to him. Its bright bluish-white head and long ruddy tail—like a flame seen through smoke, he said—persisted through autumn into winter. That lengthy visitation gave Tycho time to prove that comets, though generally assumed to be quirks of the Earth's atmosphere, actually traced paths among the planets. In contrast to contemporaries who feared the comet augured famine and pestilence, maybe even the death of a leader, Tycho confined its wrath to the heavens themselves. The Great Comet of 1577 condemned the ancient notion that solid celestial spheres carried the planets on their eternal rounds. Tycho saw plainly that no such structures impeded the comet's free travel, and therefore concluded that no such structures existed. When he delivered this thunderbolt, one could almost hear the tinkle of shattering crystal.*

* While working together in Frauenburg, Rheticus and Copernicus observed a comet that they judged to be supralunar, just as Tycho later demonstrated to the world. Rheticus wrote to his friend Paul Eber about their discovery, and Eber in turn reported it to Melanchthon in a surviving letter of April 15, 1541.

Tycho Brahe, Lord of Uraniborg.

Tycho's admittedly nonacademic achievements soon gained him an adjunct faculty position at the University of Copenhagen, where he lectured briefly on Copernicus's ideas and distributed the *Prutenic Tables* to his students. In addition to having read *On the Revolutions*, Tycho also acquired a handwritten copy of the *Brief Sketch* from a friend who had known Rheticus. Recognizing the mathematical importance of the document, Tycho made additional copies to distribute among other mathematicians, though he refused to accept the reality of the Earth's motion. Bold as he was, and openly admiring of Copernicus, he stood firm on the stationary Earth. For, if the Earth truly pursued a great circle around the Sun, Tycho reasoned, then an Earthly observer would see the spaces between certain stars widen and narrow over the course of the year. He estimated the expected change, called parallax, at 7°, or about fifteen times the diameter of the full Moon. Tycho's failure to perceive any parallax, even a tiny one, convinced him that no Earthly revolution took place. Copernicus's explanation—that the stars' tremendous distance precluded the perception of parallax—rang hollow to Tycho. Why, he asked, should the distance to the stars mushroom from Ptolemy's ten thousand Earth diameters to the several million required by Copernicus? What purpose would all that emptiness serve? What's more, stars visible across such immense gulfs would need to be absurdly large, perhaps bigger than the entire expanse of Copernicus's great circle. Incredulous, Tycho sought alternative means to realize the best of Copernicus's ideas without moving the Earth, and came up with the compromise that bears his name. In the Tychonic system, Mercury, Venus, Mars, Jupiter, and Saturn all orbit the Sun, while the Sun, in turn, carries them along as it orbits the central, immobile Earth.

In order to prove the superiority of his system, published in 1588, over the Ptolemaic or the Copernican, Tycho needed reliable data—such data as had never before been available—regarding the planets' motions. He single-handedly set new standards for accuracy and precision in observation, first by expanding the sizes of his custom-made instruments to giant proportions. In place of a handheld cross-staff or pair of compasses, for example, Tycho substituted a mammoth quadrant that stood twenty feet high and required a crew of servants to operate. Later he fashioned other devices—still grand but not quite so unwieldy—that yielded good readings on large, legible scales, where each degree of arc divided into its full complement of sixty minutes (and in some cases further subdivided into multiples of arc-seconds). With the cooperation of his prestigious family, he built his country's first astronomical observatory. King Frederick then provided the land and funding for a second one, equipped with more and still grander tools of Tycho's design, which proved, by all accounts, the finest instruments in the world for pinpointing planetary positions. Both Tycho and his magnificent observatory, Uraniborg, on the island of Hven, drew income from canonries and other Church benefices assigned to them by the king. Here Tycho ruled a staff of talented assistants, a workforce of disgruntled peasants, and the whole of the night sky for more than twenty years.

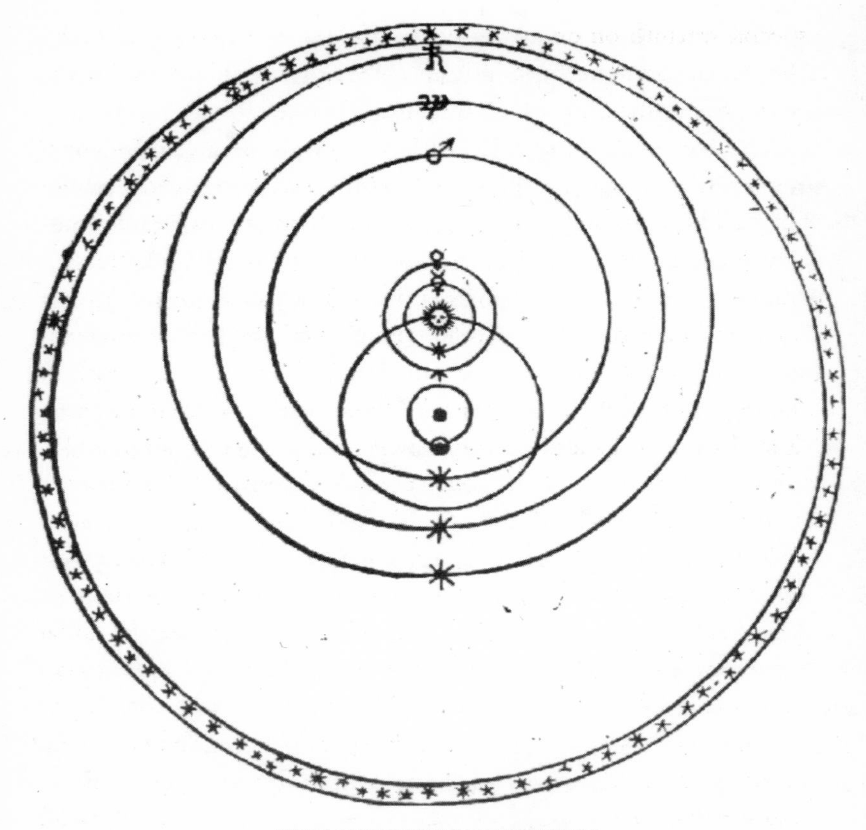

THE TYCHONIC SYSTEM

Tycho set the planets in orbit around the Sun, but left the Earth immobile at the center of the universe. Although Tycho's observations demonstrated the heavens' solid spheres to be a fiction, Tycho could not bring himself to believe the Earth rotated and revolved.

After Frederick died, Tycho fell out of favor with Christian, the heir to the Danish throne, and felt forced to abandon Uraniborg. The search for a new patron led him to Prague in 1599, to the court of the Holy Roman Emperor, Rudolf II. Although Catholic, Rudolf acted liberally toward Lutherans in general, and smiled with special warmth on one so skilled in the art of astrology as Tycho Brahe. The emperor gave him his choice of castles and put him to work prognosticating affairs of state.

The move to Prague also put Tycho in proximity to Johannes Kepler, thereby facilitating their fateful collaboration. Kepler, not yet well known to most astronomers and living in modest circumstances, could never have afforded a visit to Tycho's island. He welcomed Tycho's presence in Bohemia as an act of God. By further provision of Providence, Kepler found Tycho's chief assistant engaged in Mars studies when he joined the team at Benatky Castle in the spring of 1600. "I consider it a divine decree," reflected Kepler, "that I came at exactly the time when he was intent upon Mars, whose motions provide the only possible access to the hidden secrets of astronomy."

Kepler had been an infant in arms in Weil der Stadt in southwest Germany the year of Tycho's nova, but when he was five, his mother took his hand and led him up a hill outside town to see the Great Comet of 1577. By then, Kepler's eyesight had already begun to fail him. Likewise the noble standing that once elevated the Kepler family name had eroded before his birth, so that the myopic young genius inherited little more than a coat of arms. By dint of his intellect, however, he won scholarships that carried him all the way through seminary and university. He focused his self-professed "burning eagerness" on studies of astronomy that convinced him of the correctness of the Copernican hypothesis.

Although Kepler prepared himself for a career as a Lutheran pastor, he accepted the first job offer he received—that of a secondary school teacher and district mathematician in a provincial outpost. One day in 1595 while at the blackboard, sketching the repetition pattern of Jupiter-Saturn conjunctions for his class, he experienced an epiphany. Geometry and divinity combined in his mind, helping him intuit the solution to three cosmic mysteries: why the planets assumed their specific distances from one another, why God had created only six of them, and why they revolved at different speeds around the Sun. In the first moments of excitement, Kepler envisioned the spheres of the planets as though each were inscribed inside a particular form of regular polygon—from triangle to square, pentagon, hexagon, and so on. But, since there were any number of regular polygons and only six planets, Kepler soon scrapped them for their rarer three-dimensional counterparts, called regular solids. The simplest of these, the tetrahedron, with four faces made by four identical equilateral triangles, fitted handily between the spheres of Mars and Jupiter. The cube (comprising six equal squares) accounted for the distance between Jupiter and Saturn, and the dodecahedron (consisting of twelve cloned pentagons) accommodated the Earth within the orbit of Mars. The glorious confluence of the five regular solids with the five interplanetary interstices flooded Kepler's soul. It brought him to tears and redirected all his efforts.

"Days and nights I passed in calculating," he reported in his 1596 book, *Mysterium cosmographicum*, "to see whether this idea would agree with the Copernican orbits, or if my happiness would be carried away by the wind." At length he made everything fit. But he craved further confirmation, and that, Kepler knew, could come only from Tycho—from the trove of observational data collected over decades with scrupulous attention to detail.

Tycho had need of Kepler, too—of the German mathematician's superlative ability to mine the data for its hidden wealth. If, as Tycho believed, his data verified his version of cosmic order, then his life work would surpass that of Ptolemy and reward his every sacrifice. But Tycho worried that Kepler, a confessed Copernican who had reprinted Rheticus's *First Account* as an appendix to his own *Mysterium*, might uncover incontrovertible evidence for the rival theory—or that he might twist evidence that favored the Tychonic system to feign support for the Copernican. Thus the distrustful Tycho dallied, making Kepler pant for every bit of data he deigned

Johannes Kepler, imperial mathematician to Rudolf II.

to release. Only after Tycho's sudden death, in October 1601, and the inevitable struggle with Tycho's heirs over access to the data, did Kepler finally take possession of Tycho's treasure and lay it at Copernicus's feet.

"I build my whole astronomy upon Copernicus's hypotheses concerning the world," Kepler proclaimed in his *Epitome of Copernican Astronomy*. He thanked Tycho for his observations, but dismissed the Tychonic system as inferior. The Earth, he averred, most assuredly moved among the planets, around the Sun, just as Copernicus maintained. But Copernicus had centered the planets' revolutions on a point near the Sun, rather than on the Sun itself. Kepler found this notion physically implausible, so he corrected it. He relocated the center of all planetary motions in the body of the Sun, and imbued the Sun with a force that spread like light through the universe, pushing the planets now faster, now slower, depending on their distance. Not only did the planets nearest the Sun outpace the farther ones, as Copernicus had remarked, but each planet periodically altered its own distance from the Sun, and changed its speed accordingly. Kepler proved the path of a planet was not a perfect circle, or any combination of perfect circles, but the slightly flattened and double-centered circle known as an ellipse, with the Sun at one focus.

Kepler's ever-so-slightly squashed Martian orbit barely deviated from perfect roundness, even though it proved more elliptical than that of any other planet. For this reason, Kepler considered Mars's path—the one most closely trailed, most richly documented by Tycho—the "only possible" route to the truths of a "New Astronomy," rooted in the laws of physics. Had he tackled Jupiter or Saturn, for example, the subtleties of the ellipse would have escaped his notice and caused insuperable problems.

"I was almost driven to madness in considering and calculating this matter," wrote Kepler of the Mars situation. "I could not find out why the planet would rather go on an elliptical orbit. Oh, ridiculous me!"

Unlike Copernicus, who never divulged his thought processes in print, Kepler shared with his readers many details of his progress and setbacks. He gushed in reliving for them the "sacred frenzy" of his ecstatic insights, and begged their commiseration for all the despair he endured. "If you are wearied by this tedious procedure," he interjected in the description of his five-year "war" on Mars, "take pity on me who carried out at least seventy trials, with the loss of much time." Sometimes he felt lost, "hesitating in doubt of how to proceed, like a man who does not know how to put together again the dismantled wheels of a machine."

Kepler knew who had written the unsigned foreword "To the Reader" in the opening pages of *On the Revolutions*. His own secondhand copy of the first edition had the name Andreas Osiander penned in, right above the offending note, by the book's original owner—a Nuremberg mathematician with ties to Rheticus and Schöner. Kepler took particular umbrage at Osiander's assertions. In 1609, when he published his *New Astronomy* calling for a science based on physical forces, he named and attacked Osiander on the verso of the title page. Now everyone would know the identity of Copernicus's anonymous apologist. "It is a most absurd fiction," Kepler railed, "that the phenomena of nature can be demonstrated by false causes. But this fiction is not in Copernicus."

By his own self-assessment, Kepler's pioneering achievement lay in "the unexpected transfer of the whole of astronomy from fictitious circles to natural causes."

Kepler demolished antiquity's perfect circles while he was still wedded to the five regular solids as space-holders between the planetary orbits—a concept he nearly realized in silver, as a cosmic fountain for his patron the Duke of Württemberg. He never found cause to relinquish that fantastical vision of universal harmony. Nor did he ever abandon his Lutheran faith, though he felt compelled more than once to move house and change employment rather than convert to Catholicism at the insistence of local authorities. These beliefs sustained him through the difficulties of his chosen work, the deaths of his first wife and eight of his twelve children, his mother's trial for witchcraft, and the outbreak of the Thirty Years' War.

"Thee, O Lord Creator, who by the light of nature arouse in us a longing for the light of grace," Kepler prayed in 1619 at the close of his book *Harmonies of the World*, "if I have been drawn into rashness by the admirable beauty of Thy works, or if I have pursued my own glory among men while engaged in a work intended for Thy glory, be merciful, be compassionate, and pardon me; and finally deign graciously to effect that these demonstrations give way to Thy glory and the salvation of souls and nowhere be an obstacle to that."

Along with legal custody of Tycho's data, Kepler assumed the title of imperial mathematician at the court of Rudolf II, and also the duty to compile new, improved astronomical tables in the emperor's name. The *Rudolfine Tables*, published in 1627, indeed proved far superior to their *Prutenic* predecessors. Whereas predictions based on earlier tables might err by as much as five degrees, and sometimes

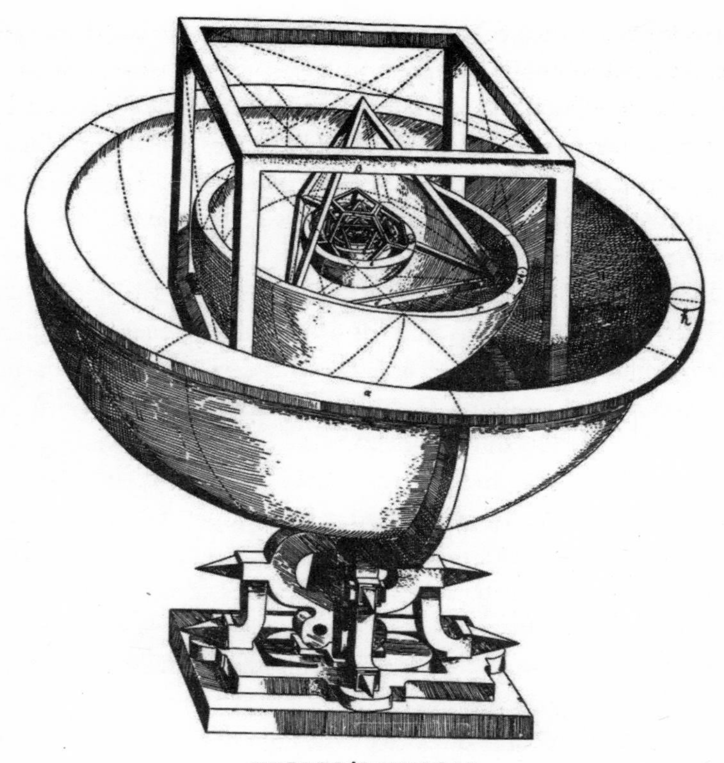

KEPLER'S VISION

While calculating conjunctions of Jupiter and Saturn, Kepler experienced a mystical vision that led him to suggest the orbits of the planets were numbered and spaced in accordance with the five regular (often called Platonic) solids. The cube that appears dominant in this image determines the interval between Jupiter and Saturn.

misjudge an important event such as a conjunction by a day or two, the *Rudolfine Tables* honed positions to within two minutes of arc.

Although the *Rudolfine Tables* trumped the *Prutenic*, Copernicus's original diagram of the Sun inside nested rings containing a planet apiece—the bull's-eye icon that he drew in his manuscript and published with *On the Revolutions*—remained strangely apt. Copernicus no doubt intended the image as a mere approximation of planetary order, since he omitted more than a score of the thirty-odd circles described in his text. But now that Tycho had swept the cosmos clean of solid spheres and Kepler banished every last epicycle, the same streamlined schematic illustration rendered an actual map of the heavens.

CHAPTER 11

Dialogue Concerning the Two Chief Systems of the World, Ptolemaic and Copernican

The constitution of the universe, I believe, may be set in first place among all natural things that can be known, for coming before all others in grandeur by reason of its universal content, it must also stand above them all in nobility as their rule and standard. Therefore if any men might claim extreme distinction in intellect above all mankind, Ptolemy and Copernicus were such men, whose gaze was thus raised on high and who philosophized about the constitution of the world.

—GALILEO GALILEI, *Dialogue Concerning the Two Chief Systems of the World*, 1632

IN AN EXCHANGE OF LETTERS in Latin between Galileo and Kepler in 1597, prompted by the publication of Kepler's *Mysterium*, the Italian Catholic professor admitted to having long been "a secret Copernican" who could not openly espouse his belief in a moving Earth for fear of ridicule from colleagues. In reply, the German Lutheran exhorted him to join the pro-Copernican movement: "Would it not be better to pull the rolling wagon to its destination with united effort?"

Galileo answered Kepler with silence. Not until 1610, after refining the optical instrument he called a spyglass, and discovering through its lenses heavenly marvels such as the moons of Jupiter, did he outwardly avow his support for Copernicus.

Before Galileo's innovations refined the rudimentary spyglass, in-

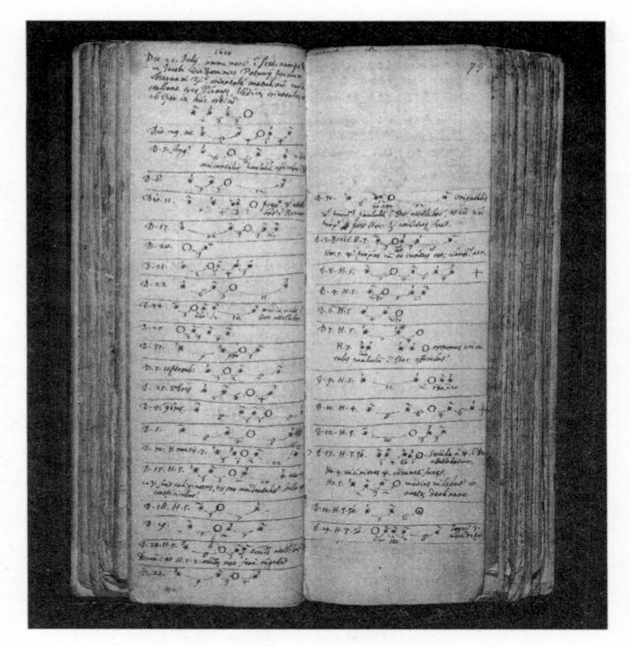

Galileo's telescopic discovery of the four largest moons of Jupiter in January 1610, described and diagrammed here in his own hand, convinced him that the Earth was not the only center of motion in the universe, and he became an outspoken "Copernican."

struments had aided astronomers in defining only the positions of bodies. Galileo's telescopes enabled observers to glimpse composition as well. The lunar landscape, for example, erupted in rocky mountains and fell into deep valleys, mirroring the surface of the Earth. The Sun exhaled dark spots that gathered and glided across its face like windblown clouds. The telescope further upset the balance of the heavens by exposing unknown bodies—not "new" entities such as Tycho's nova of 1572 (or Kepler's of 1604) that suddenly caught the naked eye, but never-before-seen objects beyond the reach of human vision, including ear-like appendages flanking Saturn and hundreds of faint stars filling in the outlines of the constellations. Also the planet Venus revealed a cycle of phases, from crescent to full, demonstrating beyond doubt that it revolved around the Sun. Venus's phases fit equally well into the Tychonic system or the Copernican, but Ptolemy's universe could not embrace such a phenomenon. Galileo published his findings. *The Starry Messenger*, a slim volume expounding his "message from the stars," sold out within one week of its printing in Padua in March of 1610. After that, he could not build telescopes fast enough to keep pace with the demand.

News of the new discoveries spread quickly to tremendous acclaim, but Galileo also became a lightning rod for all the criticism, ridicule, and outrage that Copernicus had dreaded. Thanks in part to Galileo's loud praise of it, *On the Revolutions* came to the suspicious attention of the Sacred Congregation of the Index, a watchdog arm of the Church created late in the sixteenth century to proscribe books that threatened faith or morals.

Copernicus had predicted trouble from "babblers who claim to be judges of astronomy although completely ignorant of the subject," who would distort passages of Scripture to censure him. Rheticus had also anticipated such calumny, and tried to ward it off by rectifying tenets of the Copernican system with chapter and verse, with Bishop Giese's wholehearted approval. Even Osiander, whose anonymous note "To the Reader" had so offended Giese and Kepler, probably intended to protect the book by dismissing Copernicus's bold assertions as clever calculation devices. And, as expected, *On the Revolutions* provoked the ire of religious authorities almost from the moment it appeared.

Pope Paul III, the dedicatee, had established the Roman Holy Office of the Inquisition in 1542, a year before the book's publication, as part of his campaign to quash the Lutheran heresy. Whether through the efforts of Rheticus or Giese, His Holiness duly received a gift copy of *On the Revolutions*. He turned it over to his personal theologian, Bartolomeo Spina of Pisa, Master of the Sacred and Apostolic Palace. Spina took sick, however, and died before he could review the book, leaving that task to his friend and fellow Dominican friar Giovanni Maria Tolosani. In an appendix to the treatise *On the Truth of Holy Scripture*, published in 1544, Tolosani denounced the deceased Copernicus as a braggart and a fool who risked straying from the faith.

"Summon men educated in all the sciences, and let them read Copernicus, Book I, on the moving Earth and the motionless starry heaven," Tolosani challenged. "Surely they will find that his arguments have no solidity and can be very easily refuted. For it is stupid to contradict a belief accepted by everyone over a very long time for extremely strong reasons, unless the naysayer uses more powerful and incontrovertible proofs, and completely rebuts the opposed reasoning. Copernicus does not do this at all."

Thus panned, *On the Revolutions* eluded official denunciation for a time. All works by Rheticus, however, along with those of Martin Luther, Johann Schöner, and numerous other Protestant authors, took their places on the Roman *Index of Prohibited Books* in 1559. Petreius's name turned up that same year on the appended list of forbidden printers, prompting some unknown number of zealots to destroy their copies of *On the Revolutions* because of its association with a prohibited press. Fortunately, the *Index* of 1564 reversed the situation by removing Petreius's name. Two years later, when his relative Petri issued the Basel edition, a few Catholic readers obediently excised its bonus text of the *First Account* with scissors or knives. Some also deleted Rheticus's name from that edition's title page, by crossing it out or pasting a slip of paper over it.

In Protestant regions, where the *Index* carried no weight, *On the Revolutions* nevertheless stood open to attack on religious grounds. Kepler therefore defended the Copernican idea in the introduction to his 1609 *New Astronomy*. He argued that the Holy Scriptures spoke by turns colloquially and poetically about common things such as the Sun's apparent motion through the sky—"concerning which it is not their purpose to instruct humanity." Given the Bible's emphasis on salvation, Kepler advised readers to "regard the Holy Spirit as a divine messenger, and refrain from wantonly dragging Him into physics class."

Galileo seconded Kepler's stand on biblical interpretation. "I believe that the intention of Holy Writ was to persuade men of the truths necessary for salvation," he wrote in a position paper in 1613, "such as neither science nor any other means could render credible, but only the voice of the Holy Spirit. But I do not think it necessary to believe that the same God who gave us our senses, our speech, our intellect, would have put aside the use of these, to teach us instead such things as with their help we could find out for ourselves, particularly in the case of these sciences of which there is not the smallest mention in the Scriptures; and, above all, in astronomy, of which so little notice is taken that the names of none of the planets are mentioned. Surely if the intention of the sacred scribes had been to teach the people astronomy, they would not have passed over the subject so completely."

Galileo greatly expanded his comments two years later, in 1615, in reaction to rumors that the Holy Office planned to list *On the Revolutions* on the *Index*. Addressing himself to the Medici matriarch, the Grand Duchess Cristina, he pointed out the folly of such an action:

"To ban Copernicus now that his doctrine is daily reinforced by many new observations and by the learned applying themselves to the reading of his book, after this opinion has been allowed and tolerated for those many years during which it was less followed and less confirmed, would seem in my judgment to be a contravention of truth, and an attempt to hide and suppress her the more as she revealed herself the more clearly and plainly. Not to abolish and censure his whole book, but only to condemn as erroneous this particular proposition, would (if I am not mistaken) be a still greater detriment to the minds of men, since it would afford them occasion to see a proposition proved that it was heresy to believe. And to

Galileo Galilei, philosopher and mathematician to the Grand Duke of Tuscany, in a drawing by Ottavio Leoni.

prohibit the whole science would be but to censure a hundred passages of Holy Scripture which teach us that the glory and greatness of Almighty God are marvelously discerned in all His works and divinely read in the open book of Heaven."

Galileo confronted Joshua head-on. He considered the miracle first from the Ptolemaic—Earth-centric, Earth-static—point of view, and then claimed the Copernican universe far better equipped to answer Joshua's prayers.

"Now let us consider the extent to which it is true that the famous passage in Joshua may be accepted without altering the literal meaning of its words, and under what conditions the day might be greatly lengthened by obedience of the Sun to Joshua's command that it stand still.

"In the Ptolemaic system, this could never happen at all. For the movement of the Sun through the ecliptic is from west to east, and hence it is opposite to the movement of the *primum mobile*, which in that system causes day and night. Therefore it is obvious that if the

Sun should cease its own proper motion, the day would become shorter, and not longer. The way to lengthen the day would be to speed up the Sun's proper motion; and to cause the Sun to remain above the horizon for some time in one place without declining towards the west, it would be necessary to hasten this motion until it was equal to that of the *primum mobile*. This would amount to accelerating the customary speed of the Sun about 360 times. Therefore if Joshua had intended his words to be taken in their pure and proper sense, he would have ordered the Sun to accelerate its own motion in such a way that the impulse from the *primum mobile* would not carry it westward. But since his words were to be heard by people who very likely knew nothing of any celestial motions beyond the great general movement from east to west, he stooped to their capacity and spoke according to their understanding, as he had no intention of teaching them the arrangement of the spheres, but merely of having them perceive the greatness of the miracle."

Galileo next considered the possibility that Joshua meant to halt the *primum mobile*, and all heavenly movement along with it. "And indeed Joshua did intend the whole system of celestial spheres to stand still, as may be deduced from his simultaneous command to the Moon, which had nothing to do with lengthening the day. And under his command to the Moon we are to understand the other planets as well, though they are passed over in silence here as elsewhere in the Bible, which was not written to teach us astronomy."

Turning to Copernicus, Galileo reminded Madama Cristina of his own discovery of the Sun's monthly rotation, which he propounded in his 1613 *Letters on Sunspots*.

"If we consider the nobility of the Sun, and the fact that it is the font of light which (as I shall conclusively prove) illuminates not only the Moon and the Earth but all the other planets, which are inherently dark, then I believe that it will not be entirely unphilosophical to say that the Sun, as the chief minister of Nature and in a certain sense the heart and soul of the universe, infuses by its own rotation not only light but also motion into other bodies which surround it. And just as if the motion of the heart should cease in an animal, all other motions of its members would also cease, so if the rotation of the Sun were to stop, the rotations of all the planets would stop too."

Thus, stopping the Sun sufficed to halt "the whole system of the world." All the heavenly revolutions ceased as a result, and "day was miraculously prolonged." For emphasis, Galileo pointed out how "exquisitely" his scenario agreed with "the literal sense of the sacred text."

Continuing with brio, Galileo brought up the matter of the Sun's having stood still "in the midst of the heavens," according to Joshua 10:13, and he elaborated on that point.

"Grave theologians raise a question about this passage, for it seems very likely that when Joshua requested the lengthening of the day, the Sun was near setting and not at the meridian. . . . For if it had been near the meridian, either it would have been needless to request a miracle, or it would have been sufficient merely to have prayed for some retardation." This conundrum had forced several biblical scholars, named here by Galileo, to equivocate in their interpretation of the phrase "in the midst of the heavens." But all inconsistencies vanished "if, in agreement with the Copernican system, we place the Sun in the 'midst'—that is, in the center—of the celestial orbs and planetary rotations, as it is most necessary to do. Then take any hour of the day, either noon, or any hour as close to evening as you please, and the day would be lengthened and all the celestial revolutions stopped by the Sun's standing still *in the midst of the heavens*; that is, in the center, where it resides."

In his enthusiasm for the Copernican system, Galileo apparently forgot that Catholic law forbade such exegesis by a layman. Only the Holy Fathers of the Church were empowered to probe the deep meaning of the Bible. The Protestant Kepler could emulate Luther and reach a personal understanding of Holy Writ in his own country with impunity. But Galileo, according to the decrees of the Council of Trent, issued in 1564, dared not interpret Scripture "in any way other than in accordance with the unanimous agreement of the Fathers."

The Fathers counted saints and martyrs of old among their number, and also the cardinal inquisitors of Galileo's day, including the pope's Jesuit theological adviser, Roberto Bellarmino, who put down Galileo's arguments with his own authoritative declaration:

"The words 'the Sun also riseth and the Sun goeth down, and hasteth to his place where he arose, etc.' were those of Solomon, who not only spoke by divine inspiration but was a man wise above all others and most learned in human sciences and in the knowledge of all created things, and his wisdom was from God. Thus it is not likely that he would affirm something which was contrary to a truth either already demonstrated, or likely to be demonstrated. And if you tell me that Solomon spoke only according to the appearances, and that it seems to us that the Sun goes around when actually it is the Earth which moves, as it seems to one on a ship that the shore moves away from the ship, I shall answer that though it may appear to a voyager as if the shore were receding from the vessel on which he stands, rather than the vessel from the shore, yet he knows this to be an illusion and is able to correct it because he sees clearly that it is the ship and not the shore that is in movement. But as to the Sun and the Earth, a wise man has no need to correct his judgment, for his experience tells him plainly that the Earth is standing still and that his eyes are not deceived when they report that the Sun, Moon, and stars are in motion."

On February 23, 1616, a panel of eleven theologians put the Copernican idea to a vote. They deemed "the quiescence of the Sun in the center of the world" to be "formally heretical" because it contradicted Scripture. They further found the heliocentric universe philosophically "foolish and absurd." Although the Earth's motion seemed to them an equally ridiculous concept, they declared it merely "erroneous in faith," since it did not explicitly deny the truth of Holy Writ. These judgments formed the crux of an official edict issued on March 5, denouncing Copernicus's teachings as "false and contrary to Holy Scripture." *On the Revolutions* would be named in a decree appended to the *Index of Prohibited Books*. But instead of being banned and destroyed—the fate of other forbidden titles—*On the Revolutions* was to be suspended until corrected. In the decades since its publication, the book had proved so useful that the Church could not justify condemning it outright. Indeed, the much-desired calendar reform that engaged Copernicus during his lifetime had since been implemented with the help of his text. *On the Revolutions* and the *Prutenic Tables* provided the mean length of the tropical year and synodic month that enabled the Jesuit Father Christoph Clavius of the Roman College to create the so-called Gregorian calendar, which replaced the Julian in 1582, during the pontificate of Gregory XIII.

In 1619 another decree regarding the *Index* banned Kepler's *Epitome of Copernican Astronomy*, along with "all other works by this author." The following year a further decree enumerated ten specific corrections to be made to *On the Revolutions*. These few changes—only ten points in more than four hundred pages—conformed Copernicus's text to Osiander's note. They rephrased every testimony to the Earth's motion so as to sound hypothetical. The censors deleted the part of the preface claiming "astronomy is written for astronomers," for they had appropriated the subject to themselves. The line appeared in the paragraph that voiced Copernicus's nightmare vision of "babblers who claim to be judges of astronomy although completely ignorant of the subject," who might bend "some passage of Scripture to their purpose" and bludgeon him with it.

It was left to each individual owner of *On the Revolutions* to insert the specified changes where indicated. Galileo made all of them, perhaps in the belief that his copy might be examined by clerical authorities. He had himself been advised personally by Cardinal Bellarmino, in Rome in 1616, to end his teaching and writing about

In his best-known work, Dialogue Concerning the Two Chief Systems of the World, *Galileo staged a four-day conversation among three intellectuals. The frontispiece to the first edition pictured those men as Aristotle, Ptolemy, and Copernicus (at right, holding a representation of his Sun-centered cosmos).*

Copernicus, and he had acquiesced. Several years later, however, in 1624, a seemingly broad-minded new pope, Urban VIII, encouraged Galileo to write a definitive comparison of the Ptolemaic system with the Copernican. This book, Galileo's *Dialogue Concerning the Two Chief Systems of the World,* published in Florence in 1632, led to immediate accusations of heresy. The formal trial of Galileo by the Inquisition took place the following year, and ended with his abjuration. The *Dialogue* then joined *On the Revolutions* on the next *Index of Prohibited Books*. They both remained listed there—the one banned, the other suspended until corrected, both the subjects of continuing controversy and commentary—through the ensuing two centuries.

Chapter 12

An Annotated Census of Copernicus' De Revolutionibus

I have compared the editions of 1543, 1566, and 1617. The latter is the best of the three, that of 1566 only a reprint of the editio princeps *of 1543. There is only one difference: that of 1543 has an ample errata . . . which disappears in 1566 without any of the errors being corrected. This is all corrected in 1617 and moreover the text has been rectified in many places; also there are notes by the editor and some very useful examples of calculations. Book lovers can prefer the* editio princeps, *which is on beautiful paper and which is larger and has more body, with letters on the figures less tiny, but astronomers must use the edition of 1617. This is why I keep two examples. I had the edition of 1566, but gave it up.*

—J. B. J. Delambre, director of the Paris Observatory, 1804–1822, from a comment tacked inside a 1566 copy of *On the Revolutions*

he condemnation of Copernicus's ideas by the Roman Church, which would have devastated the Catholic canon had he lived to hear of it, probably served to make his book more popular. Only one year after the Edict of 1616 connected his name with the crime of heresy, a third edition of *On the Revolutions* appeared in Amsterdam, under the title *Astronomy Renewed*. The printer was Willem Janszoon Blaeu, a revered cartographer and globe maker who had apprenticed for six months under Tycho on the island of Hven. Nicolaas Müller, a professor of both medicine and mathematics at the University of Groningen in the Netherlands, wrote notes and commentary for this edition, seventy-four years after Copernicus's death. Numerous commemorative editions followed, including the 1873 in Torun, to celebrate four hundred years since the author's birth there. From Turin, Italy, in 1943, came a first-edition facsimile on the occasion of the book's quadricentennial. The 1973 Copernicus quinquecentennial inspired a facsimile of the original manuscript, in two colors of ink on aged-looking deckled pages, reproducing even the accidental smudges and blots. The facsimile lacks only a single detail—the small hole that Copernicus drilled accidentally in his original, by anchoring one leg of his compasses in the same spot, eight times over, to execute the concentric circles for his diagram of the heavenly spheres.

In 1970, while planning how to pay homage to Copernicus for his five hundredth year, Owen Gingerich of Harvard University conceived a unique research project. He had been combing through a cache of rare astronomy books at the Royal Observatory in Edinburgh, Scotland, when he turned up a first edition of *On the Revolutions*. Unlike the few other copies he had seen—and much to his surprise—this one proved "richly annotated from beginning to end." Gingerich had read Arthur Koestler's history of early astronomy, *The Sleepwalkers*, in which *On the Revolutions* was dismissed as "the book that nobody read." But the Edinburgh copy had obviously been read with care—studied, in fact. Between the initials stamped on the book's binding and the distinctive handwriting in the glosses that filled the margins, Gingerich soon identified the original owner as Erasmus Reinhold, senior mathematician at the University of Wittenberg. This very volume of *On the Revolutions* had conjured the *Prutenic Tables*. Gingerich wondered how many other significantly scribbled-in copies might be hiding in scholarly libraries and private collections around the world. What he came to call his "Great Copernicus Chase" occupied him for the next thirty years. *An Annotated Census of Copernicus'* De Revolutionibus, published in 2002, presents Gingerich's hard-won descriptions of 277 copies of the first edition and 324 of the second. After finding Reinhold's copy, Kepler's copy, Galileo's, and more, Gingerich could safely say that *everybody* read *On the Revolutions*.

If someone were to gather the six-hundred-plus surviving sixteenth-century copies together for a grand reunion in a great exhibition hall, the books would not resemble one another much at all, despite their shared title and text. Because books printed in the sixteenth century were generally sold unbound, each copy is distinguished by its owner's choice of binding, from expensive calf to cheaper varieties of sheepskin—most often in white, but also black, red, bright red, brown, tan, slate, gray, green, yellow, and orange—laid over boards of oak or cardboard, and stamped with initials or coats of arms in gold, or scenes from the Bible, or saints, or medallions of Martin Luther and Philip Melanchthon. Many such covers once shut tight with ornately carved clasps, or with leather or colored silk ties dyed to match the speckled edges of the pages and complementing some hue in the marbled endpapers. Even its earliest owners considered *On the Revolutions* a big, important book, well worth some trouble and expense. As Gingerich discovered from the prices noted in a few copies, scholars paid as much as seventeen *groschen* for it in Wittenberg and Stockholm, roughly double the cost of a university semester's matriculation fee.

The wide margins of the pages tell another story altogether—a chronicle of interactive education, in which the new astronomy passed from hand to hand and generation to generation. Some copies contain annotations in two or even three different hands, and in several cases whole series of nearly identical notes repeat in numerous volumes, demonstrating the influence of certain teachers.

Among the most extensively and tellingly annotated copies is the one that belonged to Johannes Kepler, now held at the Universitätsbibliothek in Leipzig. It is a first edition, first owned by Jerome Schreiber of Nuremberg, who received it as a gift from the printer. Petreius's personalized inscription can still be read on the title page (where someone has crossed out the words "of the Heavenly Spheres"). Presumably the two men knew each other through Johann Schöner, who tutored Schreiber in mathematics. Or through Rheticus, who was Schreiber's classmate at Wittenberg and later his colleague on the faculty. Over the four years Schreiber owned the book, until his death at age thirty-two, he wrote copious notes in it. He corrected every typographical error stipulated on the errata leaf, as well as those in the remaining fifty folios beyond the scope of the errata leaf. These are the same changes that appear in Copernicus's hand in the original manuscript. Copernicus must have conveyed his final edits too late for Petreius to include them, but still in time for Rheticus to share them with a small circle of friends. Gingerich found only nine copies of *On the Revolutions* so thoroughly set right.

It was Schreiber, a true insider, who knowingly penned the name Andreas Osiander above the anonymous note to the reader.

Schreiber also copied Rheticus's marginal notes into his own copy, and mused alongside the text about questions he wished he could have asked Copernicus. On folio 96, for example, Copernicus waffled on whether the center of the universe lay within the Sun or at the empty center of the Earth's orbit. He said he would take up the matter later, but then never got back to it. Schreiber noted here that Rheticus had decided the point, in the *First Account*, in favor of the Sun as center. On folio 143, next to Copernicus's concession that his use of a small epicyclet resulted in an orbit with a noncircular shape, Schreiber jotted a single word, in Greek. Kepler, too, could read and write Greek, and so, when he purchased Schreiber's copy in 1598 and came across the word ἐλλειψη at the end of Book III, chapter 25, he knew that it meant "ellipse."

"It is very remarkable," writes Gingerich, fairly shouting in his quiet way, "that of all the possible copies of the book that he might have acquired, Kepler got one with the word *ellipse* written in the margin by a highlighted passage."

The binding, in calf with a gilt-decorated spine, reflects the book

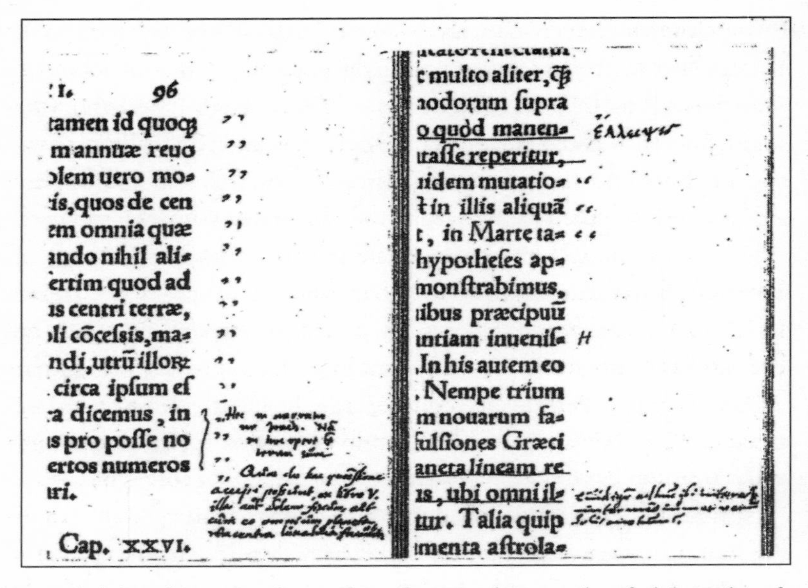
I. 96
tamen id quoq;
m annuæ reuo
olem uero mo-
is,quos de cen
em omnia quæ
ando nihil ali-
ertim quod ad
is centri terræ,
oli cōcessis,ma-
ndi,utrū illorū
circa ipsum es
a dicemus, in
is pro posse no
ertos numeros
ri.
Cap. XXVI.

c multo aliter,q̃
nodorum supra
o quód manen-
trasse reperitur,
idem mutatio-
t in illis aliquā
t, in Marte ta-
hypotheses ap-
monstrabimus,
ibus præcipuū
ntiam inuenis-
,In his autem eo
.Nempe trium
m nouarum fa-
fulsiones Græci
aneta lineam re
is, ubi omni il-
tur. Talia quip
menta astrola-

Kepler's heavily annotated copy of On the Revolutions *identified the author of the anonymous note to the reader, and also contained the word* ellipse *written as a marginal gloss in Greek.*

fashions of the eighteenth century—clearly the choice of a subsequent owner. That unidentified individual took pains, however, to preserve Kepler's annotations, by instructing the bookbinder to fold in a few selected pages rather then trim them all by the five millimeters required for a neat new finish. Unfolded, those pages display Kepler's principal comments, in the form of questions, concerning the true center of planetary motions and the need for Earth to vary its orbital speed in the same manner as the other planets.

Referring to a lettered diagram, in which Copernicus had labeled the center of the Earth's orbit as *D*, Kepler asked, "For is the orbit really such that D is the Sun, or of a temporary circle wherein D is other than the Sun?" The minuscule scale of Kepler's writing suggests he leaned in close enough for his nose to almost touch the pen nib. "Does the Earth have a simple or double difference?

Therefore are they [Venus and Mercury] nonetheless attached to the irregularities of the Earth? For the very reason, to be sure, that they are moved in other circles than the one eccentric to the Earth (that is, in concentrics and epicycles), they effect this present irregularity."

Gingerich points out that Kepler's notes, though sparse, crop up at critical places, uncovering fundamental flaws in Copernicus's theories. And although Kepler could be trusted to locate those points without help from anyone, nevertheless a trail was marked for him.

If Isaac Newton owned a copy of *On the Revolutions*, it has not survived. In his student days, he undoubtedly consulted one of the three Trinity College first editions still held by that venerable library. After Newton established universal gravitation as the force that kept the planets in their orbits around the Sun, copies of Copernicus's book came into the possession of many other giants in astronomy, such as comet namesake Edmond Halley, his successor as astronomer royal George Biddell Airy, computing pioneer Charles Babbage, and twentieth-century cosmologist Edwin Hubble, who was first to appreciate the infinite extent and continuing expansion of the universe.

Now that Copernicus's text no longer serves to describe the known paths of the planets, it is more highly valued than ever as an icon. The most recent copy of the book offered at auction—a clean, unannotated first edition—sold at Christie's, New York, in June 2008 (to an undisclosed recipient) for $2,210,500.

The *First Account*, according to Gingerich, is ten times rarer a find than *On the Revolutions*. While compiling the *Annotated Census*, he came across thirty-seven copies of Rheticus's book, the majority of them in Germany. In 2004, the Linda Hall Library of Science, Engineering & Technology in Kansas City, Missouri, bought a copy of the *First Account* for $1.5 million. The library has since digitized the entire volume, enabling anyone with an Internet connection to page through it. First-edition copies of *On the Revolutions* are also available for perusal on several rare-book Web sites.

Copernicus the man has gained iconic status as well. Statues of him proliferate, especially in Poland, where his image has frequently appeared on stamps, coins, and banknotes. His very bones became the goal of an archaeological dig begun in the summer of 2004 under the stone floor of the Frombork Cathedral, where searchers eventually unearthed the skull and several bones of a seventy-year-old man that seemed to answer his description. The skull was just a fragment—the cranium without the mandible—but its age and resting place near the altar of St. Wenceslaus (now called Holy Cross) provided strong clues. Police forensic artists, accustomed to portraiture based on partial descriptions, parlayed the chinless skull into a full face with a big, broken nose and jutting square jaw. After a perpetual Copernican youth fostered by the single image of him in his prime, the sudden weight of years distorted the astronomer's looks beyond recognition. In the photo released to news services, the old man wore a fur-collared jacket in a red reminiscent of his portrait jerkin.

Subsequent examination of the skull suggests that the dent over the right eye is an arterial depression typical of many skeletons—not a match for the scar depicted in Copernicus's portrait. No one doubts, however, that the skull belonged to him.

The scant bones underwent DNA analysis, for anticipated comparison with the latter-day descendants of Copernicus's nieces. The most convincing piece of evidence emerged from a secondary trove of remains—the nine hairs that had worked their way into Copernicus's oft-consulted copy of a 1518 calendar of eclipse predictions, held in Uppsala with the rest of the books the Swedish Army took from the Varmia library during the Thirty Years' War.

When Gingerich heard about the hairs found in the *Calendarium Romanum Magnum*, he imagined they might belong to him, given the number of times he had bent his own head over that same book to study Copernicus's notations in it. But DNA testing of the four suitable hairs showed that two of them made convincing matches

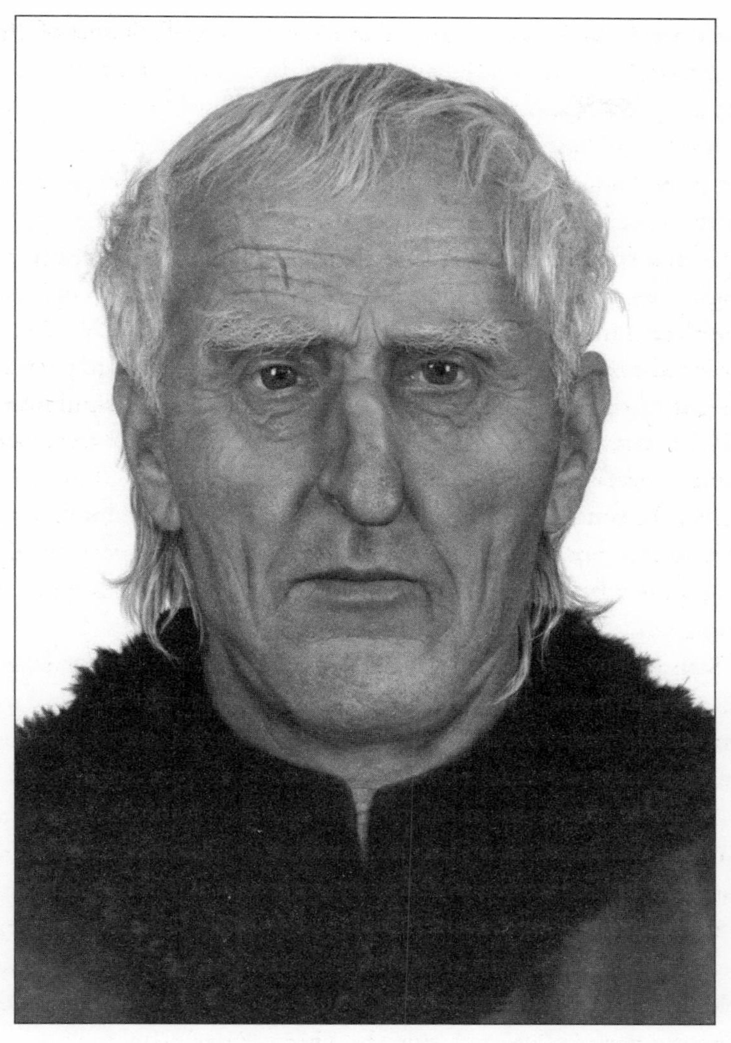

Working from bone fragments, Polish forensic artists imagined the face of their famous countryman as a man of seventy years—Copernicus's age at death.

with a well-preserved molar in the Frombork skull. When the scientists reported their results in 2009 in the *Proceedings of the National Academy of Sciences*, they said that certain genes seen in the remains were typical of blue-eyed individuals. The police photo had depicted Copernicus with brown eyes, the same color as in the Torun portrait, adjusted for age-related fading.

The second burial of Canon Nicolaus Copernicus took place in Frombork on May 22, 2010. Unlike the first funeral, the pageantry of this one drew a large crowd, and included a Mass led by the Primate of Poland, the country's most highly honored bishop. Not since Copernicus's Uncle Lukasz Watzenrode carried St. George's head here in 1510 had this cathedral seen a more triumphant procession focused on funerary relics.

A new black granite tombstone with a stylized golden Sun and planets now flanks the earlier memorial installed in the cathedral in 1735 (to replace a still earlier epitaph destroyed during wartime). Beyond such plaques, statues, and other public tributes, his fellow "mathematicians" continue to afford Copernicus their professional recognition. The first cartographers of the Moon named a large lunar crater for him in the 1600s, and Space Age explorers launched an orbiting astronomical observatory called *Copernicus* in 1972. For the 537th return of his birthday, on February 19, 2010, the International Union of Pure and Applied Chemistry announced the naming of super-heavy atomic element number 112 "copernicium" (symbol Cn) in his honor.

Every time the *Kepler* spacecraft, currently in orbit, detects a new exoplanet around a star beyond the Sun, another ripple of the Copernican Revolution reverberates through space. But the counterrevolution that sprang up in immediate reaction to Copernicus's ideas also continues to make waves. State and local governments still claim the right to control what can be taught of scientific theories in classrooms and textbooks. A so-called museum in the southeastern United States compresses the Earth's geological record from

4.5 billion to a biblical few thousand years, and pretends that dinosaurs coexisted with human beings.

Copernicus strove to restore astronomy to a prior, purer simplicity—a geometric Garden of Eden. He sacrificed the Earth's stability to that vision, and pushed the stars out of his way. To contemporaries who doubted the grandiose dimensions of the heliocentric design, Copernicus replied, "So vast, without any question, is the divine handiwork of the most excellent Almighty."

In the century after his death, the Inquisition struck that line from his text. Although Copernicus clearly meant to express confidence in the Omnipotent's ability to transcend ordinary proportion, the censors saw the statement as an ungrounded confirmation for an Earth in motion.

When the Earth moved despite the Church's objections, Copernicus

A small corner of today's known universe, depicted in this Hubble telescope Deep Field image, is many times more vast than the once-shocking distance Copernicus allowed between Saturn and the stars. To use his word, the extent of his entire cosmos was "negligible" compared with the many millions of light years separating our Milky Way from the galaxies beyond.

became symbolic of a new fall from grace. Because of him, humanity lost its place at the center of the universe. He had initiated a cascade of diminishments: The Earth is merely one of several planets in orbit around the Sun. The Sun is only one star among two hundred billion in the Milky Way—and relegated to a remote region far from the galactic center. The Milky Way is just one galaxy in a Local Group of neighbors, surrounded by countless other galaxy groups stretched across the universe. All the shining stars of all the galaxies are as nothing compared to the great volume of unseen dark matter that holds them in gravitational embraces. Even dark matter is dwarfed by the still more elusive entity, dark energy, that accounts for three quarters of a cosmos in which the very notion of a center no longer makes any sense.

THANKSGIVING

It would be impossible to overstate the generosity of the historians, directors, and people of good cheer who have helped me relive the Copernican Revolution.

Professors Owen Gingerich, André Goddu, Michael Shank, Noel Swerdlow, and the late Ernan McMullin lent their authority in history and astronomy, along with earnest encouragement. The first three also reviewed the draft chapters of this book to correct my mistakes.

Directors Gerald Freedman, Langdon Brown, and Isaac Klein read and commented on numerous drafts of the play, always with constructive advice.

In Poland, Janusz Gil, Tomasz Mazur, Krzysztof Ostrowski, and Jaroslaw Wlodarczyk variously welcomed, guided, mentored, interpreted, and read drafts for me. I am also grateful to Stanislaw Waltos for facilitating permission to view the Copernicus manuscript in Krakow.

At Uppsala, Tore Frängsmyr opened the doors to Copernicus's personal library.

The John Simon Guggenheim Memorial Foundation and the Alfred P. Sloan Foundation provided grant support, with kind attention from Edward Hirsch and André Bernard at Guggenheim, Doron Weber at Sloan, Annie MacRae as the Sloan Project manager at Manhattan Theatre Club, and Paige Evans.

The Naked Stage, the Manhattan Theatre Club, the New York State Writers Institute Authors Theatre, and the University and

Teatr Lubuski of Zielona Gora (Poland) staged readings of the play in progress.

Astrologer Elaine Peterson cast and interpreted horoscopes for Copernicus and Rheticus.

Stalwart supporters including my agent, Michael Carlisle; my editor and publisher, George Gibson; my daughter, Zoe Klein; my brothers and sister-in-law, Stephen, Michael, and Pamela Sobel; my cousins Celia Michaels and Barry Gruber; my friends Diane Ackerman, Jane Allen, Will Andrewes, K. C. Cole, Doug Garner, Mary Giaquinto, Joanne Julian, M. G. Lord, Doug Offenhartz, Rita and Gary Reiswig, Lydia Salant, Margaret Thompson, and Alfonso Triggiani have all proved especially helpful, often just by being who they are.

COPERNICAN CHRONOLOGY

1466 Peace of Torun concludes the Thirteen Years' War between the Prussian cities of Poland and the Knights of the Teutonic Order.

1473 Copernicus born, February 19.

1484 Copernicus's father dies.

1489 Copernicus's uncle Lukasz Watzenrode elected Bishop of Varmia, February 19.

1491 Copernicus enters Jagiellonian University in Krakow.

1492 Ferdinand and Isabella expel the Spanish Jews.

Columbus voyages to the New World.

1496 Copernicus studies canon law in Bologna.

1497 Copernicus appointed canon in Frauenburg.

1500 Copernicus spends several months in Rome, gives lectures on math.

1501 Copernicus and brother, Andreas, attend Varmia Chapter meeting in July.

Copernicus enrolls as medical student at Padua in October.

1502 University of Wittenberg founded.

1503 Copernicus receives doctor of canon law degree at Ferrara, May 31; becomes bishop's secretary and personal physician at Heilsberg in the fall.

1504 Copernicus observes great conjunction in Cancer, notes that Mars is ahead of—and Saturn behind—predicted positions.

1508 Copernicus conceives the geokinetic idea, probably begins work on his heliocentric model.

1509 Copernicus publishes his Latin translation of the Greek letters of Theophylactus Simocatta in Krakow.

1510 Copernicus leaves the bishop's service, moves to Frauenburg, distributes his *Brief Sketch* as a pamphlet.

1512 King Sigismund I marries Barbara Zapolya in Krakow, February 8.

Uncle Lukasz dies in Torun, March 29, after attending the king's wedding.

1513 "Doctor Nicholas" purchases bricks and lime to build an observing platform.

1514 Georg Joachim Iserin (later Rheticus) born, February 16.

1515 Copernicus offers opinion on calendar reform to Pope Leo X.

Full text of Ptolemy's *Almagest* appears in print for the first time.

1516 Copernicus begins three-year term as administrator, November 11.

1517 Copernicus writes his *Meditata* on currency problems, August 15.

Martin Luther posts his 95 Theses in Wittenberg.

1518 Andreas dies in November.

1519 Teutonic Order invades Braunsberg, December 31.

1520 Grand Master Albrecht's troops set fire to Frauenburg, January 23.

1521 War with Teutonic Knights ends; peace treaty signed, April 5.

1522 Copernicus introduces currency reform based on his essay of 1517.

Johannes Werner publishes a collection of astronomy papers in Nuremberg.

1523 Bishop Fabian Luzjanski dies, January 30.

Copernicus serves as interim bishop through October, even after Maurycy Ferber is elected Bishop of Varmia in mid-April.

1524 Great Conjunction of Jupiter and Saturn in Pisces.

Copernicus writes *Letter Against Werner*, June 3.

1525 Treaty of Krakow dissolves Order of Teutonic Knights, establishes Duchy of Prussia under Duke Albrecht.

1526 Duke Albrecht marries Princess Dorothea of Denmark, February 12.

King Sigismund orders Protestant homes in Krakow burned; issues royal decree regarding new currency, July 17.

Bishop Ferber banishes Lutherans from Varmia, September 22.

1528 Rheticus's father convicted of swindling and beheaded.

1529 Johannes Dantiscus, Polish ambassador to Spain, becomes a canon of Varmia.

1530 Canon Dantiscus, still in Spain, chosen as Bishop of Kulm.

Confession of Augsburg establishes the articles of faith for Lutherans.

1531 Copernicus, as guardian of the chapter's counting table, writes his *Bread Tariff*.

1532 Canon/Bishop-designate Dantiscus returns to Poland from Spain.

Rheticus matriculates at Wittenberg using his mother's name, de Porris.

1533 Johannes Dantiscus installed as Bishop of Kulm, April 20.

Pope Clement VII hears Copernicus's theory described, summer.

1534 Alessandro Farnese elected Pope Paul III.

Luther Bible printed (in German) in Wittenberg.

1535 Bernard Wapowski visits Copernicus, tries to publish his almanac of planetary positions.

1536 Rheticus becomes lecturer in mathematics at Wittenberg.

Cardinal Schönberg's November 1 letter asks Copernicus to release his theory.

1537 Bishop Maurycy Ferber dies, July 1, replaced by Johannes Dantiscus.

Canon Tiedemann Giese becomes Bishop of Kulm.

Cardinal Schönberg dies, September 9.

1538 Rheticus goes to Nuremberg in autumn, meets Johann Schöner.

Pope Paul III excommunicates King Henry VIII.

1539 Canon Felix Reich dies, March 1.

Dantiscus issues new edict against Lutheran heresy in March.

Rheticus arrives in Frauenburg in May, completes the *First Account* September 23.

1540 *First Account* published in Danzig in March.

Rheticus returns briefly to Wittenberg to teach in December.

1541 Second printing of *First Account* in Basel.

Melanchthon and colleagues attempt reconciliation with Catholic Church.

Rheticus returns to Wittenberg, elected dean of Faculty of Arts in October.

Rheticus publishes *On the Sides and Angles of Triangles* by Copernicus.

1542 Rheticus's term as dean ends in April; he goes to Nuremberg.

Johannes Petreius begins printing *On the Revolutions* at his press in May.

Copernicus writes his dedication to Pope Paul III in June.

Pope Paul III establishes the Roman Holy Office of the Inquisition.

Rheticus leaves Nuremberg for Leipzig in October.

Copernicus suffers a stroke in late November or early December, is left paralyzed on right side.

1543 Printing of *On the Revolutions* concludes in April.

Crown Prince Sigismund Augustus marries Archduchess Elisabeth of Austria.

Copernicus dies, May 24.

1545 Pope Paul III convenes the Council of Trent.

1546 Martin Luther dies, February 18.

1547 Rheticus suffers a five-month spiritual crisis, moves to Zurich at year's end.

1548 Returning to Leipzig, Rheticus is elected dean of Faculty of Arts, October 13.

Bishop Dantiscus dies, October 27.

1549 Tiedemann Giese becomes Bishop of Varmia, January 25.

Rheticus's works listed on the *Index of Prohibited Books* along with those of Martin Luther and Johann Schöner.

Duke Albrecht appoints Andreas Osiander head theologian of new university in Königsberg.

1550 Bishop Giese dies, October 23.

1551 Rheticus publishes his *Canon of the Science of Triangles*.

Accused of sodomy, Rheticus flees Leipzig in April.

1554 Rheticus moves to Krakow in spring, works as a medical doctor.

1562 Copernicus's relative Jan Loitz renounces his canonry in order to marry, February 8.

1564 Decrees of Council of Trent prohibit interpretation of Scripture by laymen.

1566 Second edition of *On the Revolutions* published in Basel.

1572 Tycho Brahe observes "new star" in November, writes *De nova stella*.

1574 Rheticus dies, December 4.

1582 Pope Gregory XIII replaces the Julian calendar with the Gregorian.

1588 Tycho publishes his geo-heliocentric system.

1595 Bartholomew Pitiscus, Calvinist theologian and mathematician, composes his *Trigonometry*, which title establishes the enduring term for the science of triangles.

1596 Johannes Kepler publishes his *Mysterium cosmographicum*.

Valentin Otto publishes Rheticus's work as *Opus palatinum*, full of errors.

1604 Kepler observes a nova.

1609 Galileo observes the Moon and Milky Way with an early telescope.

1610 Galileo's discovery of Jupiter's moons, in January, convinces him Copernicus was right; he publishes *The Starry Messenger*.

1613 Pitiscus publishes new summary of Rheticus's work, *Mathematical Treasury*, in Frankfurt.

1616 *On the Revolutions* appears on the *Index of Prohibited Books*, "until corrected."

1617 Third edition of *On the Revolutions* published in Amsterdam.

1619 Kepler's books listed on the *Index*.

1620 The *Index* names corrections that must be made to *On the Revolutions*.

1627 Kepler publishes the *Rudolfine Tables*.

1632 Galileo publishes his *Dialogue Concerning the Two Chief Systems of the World*.

1772 Ignacy Krasicki, the last independent Bishop of Varmia, sees his diocese absorbed into the Kingdom of Prussia by the first partition of Poland.

1835 *On the Revolutions* and Galileo's *Dialogue* dropped from the *Index*.

1972 *Copernicus* satellite launched to study ultraviolet and X-ray sources in space.

2008 First-edition copy of *On the Revolutions* sold at auction for more than $2 million.

2010 Copernicus's remains, having been exhumed for scientific study, reburied in the cathedral at Frombork (formerly Frauenburg).

NOTES ON THE QUOTATIONS

The late Edward Rosen, professor of history of science at the City University of New York, translated all of Copernicus's works into English. Charles Glenn Wallis and A. M. Duncan also made translations of *On the Revolutions of the Heavenly Spheres*, often called by its Latin title, *De revolutionibus orbium coelestium*, or *De rev* for short. In some of the quoted passages, I have combined their translations.

Astronomer and historian Noel Swerdlow, now a visiting associate at Caltech, translated several Copernican documents, including the *Brief Sketch*, or *Commentariolus*, and printer Johann Petreius's open dedication letter to Rheticus.

CHAPTER I

p. 3 "The cricket . . . the wayfarers." Rosen, *Minor Works*, 30.

p. 4 "marvelous symmetry of the universe." *De rev*, I, 10 (Rosen, 22).

p. 7 "What could be more beautiful . . . divine ruling." *De rev*, I, Introduction (Rosen, 7; Wallis, 8).

p. 7 "Among the many . . . the stars." *De rev*, I, Introduction (Rosen, 7; Wallis, 8).

p. 9 "between the . . . night." *De rev*, IV, 27 (Wallis, 223).

p. 10 "with two stout . . . salt pork." Banville, 40–41.

p. 11 "The inns . . . spilled out." Banville, 43.

p. 11 "To produce . . . vinegar." BLTC Research, "Arnold of Villanova," http://www.general-anaesthesia.com/images/arnold-of-villanova.html.

p. 14 "O right . . . country." Rosen, *Minor Works*, 29.

p. 14 "Theophylactus . . . garden." Rosen, *Minor Works*, 29.

p. 14–15 "Among mares . . . Nature's tears." Rosen, *Minor Works*, 31.
p. 15 "Just as . . . rectified." Rosen, *Minor Works*, 29.
p. 15 "conspicuous . . . demeanor." Rosen, *Minor Works*, 27.
p. 15 "the scholar . . . principles." Rosen, *Minor Works*, 27.
p. 16 "Stroll through . . . dust." Rosen, *Minor Works*, 50.

CHAPTER 2

p. 17 "The center . . . lunar sphere." Rosen, *Treatises*, 58; Swerdlow, "Derivation," 436.
p. 18 "A manuscript . . . still." Gingerich, *The Book Nobody Read*, 36; Swerdlow, "Derivation," 431.
p. 18 Evidence that Copernicus did *not* know of Aristarchus's heliocentric theory was first published by Owen Gingerich: "Did Copernicus Owe . . . ?"
p. 20 "that most outstanding of astronomers." *De rev*, II, 14 (Rosen, 83).
p. 20 "shuddered." *De rev*, I, 4 (Wallis, 12).
p. 20 "All spheres . . . Sun." Swerdlow, "Derivation," 436.
p. 20 "What appear to us . . . other planet." Rosen, *Treatises*, 58–59.
p. 21 "headlong whirl." Swerdlow, "Derivation," 444.
p. 22 "Whatever motion . . . outermost heaven." Swerdlow, "Derivation," 436.
p. 22 "utterly ridiculous . . . think of." Toomer, 44.
p. 22 "for the sake . . . larger book." Swerdlow, "Derivation," 438.
p. 22 "Mercury runs . . . the planets." Rosen, *Treatises*, 90.
p. 23 "Compared to . . . imperceptible." Swerdlow, "Derivation," 436.
p. 23–24 "So vast . . . Almighty." *De rev*, I, 10 (Rosen, 22).
p. 26 "Doctor Nicolaus . . . work-yard." Gingerich and MacLachlan, 79.
p. 27 "The ancients . . . Vistula." *De rev*, V, 25 (quoted in Gingerich, *Eye*, 383; translation attributed to Ann Wegner).

CHAPTER 3

German historian Franz Hipler paid the first scholarly attention to these "Leases" in a Copernicus quadricentennial festschrift volume published at Braunsberg (near Frauenburg) in 1873. Ten years later, Leopold Prowe included excerpts from them in his two-volume biography of Copernicus (in German). Edward Rosen completed the English translation, quoted here, in 1985.

p. 29 "Stenzel . . . 1 horse." Rosen, *Minor Works*, 232.
p. 33 "Leasing . . . 1517." Rosen, *Minor Works*, 228.
p. 33 "He got . . . 2 horses." Rosen, *Minor Works*, 228.
p. 33 "Bartolt . . . Heironym, etc." Rosen, *Minor Works*, 231.
p. 34 "who is . . . in name." Rosen, *Minor Works*, 232.
p. 34 "4 horses . . . 1 scythe." Rosen, *Minor Works*, 233.
p. 34 "Gregor Knobel . . . grown up." Rosen, *Minor Works*, 228.
p. 34 "Hans Caluke . . . 4 May." Rosen, *Minor Works*, 235.
p. 34 "Jacob Wayner . . . overseer." Rosen, *Minor Works*, 234.
p. 35–36 "Jacob took . . . annual payment." Rosen, *Minor Works*, 234.
p. 36 "Gregor . . . thievery." Rosen, *Minor Works*, 232.
p. 36 "Petrus . . . away." Rosen, *Minor Works*, 233.
p. 37 "Jacob . . . old." Rosen, *Minor Works*, 233.
p. 37–38 "in the year . . . after sunrise." *De rev*, III, 3 (Evans, 457, n. 22).
p. 38 "Jacob has . . . brother." Rosen, *Minor Works*, 234.
p. 39 "When the money . . . Heaven." "The 95 Theses and Their Results," http://www.luther.de/en/anschlag.html.
p. 40 "Voytek . . . rental." Rosen, *Minor Works*, 234.
p. 40 "Lurenz . . . 4 parcels." Rosen, *Minor Works*, 234.
p. 40 "Stenzel . . . 33 marks." Rosen, *Minor Works*, 236.

CHAPTER 4

p. 41 "Coinage . . . magnitude." Rosen, *Minor Works*, 176–77.
p. 42 "the first . . . Scorpion." *De rev*, V, 14 (Rosen, 261).
p. 42 "changes and . . . men." Grafton, 53.
p. 43 "Most Gracious . . . Lord." Biskup and Dobrzycki, 74.
p. 43 "For we . . . care." Biskup and Dobrzycki, 74.
p. 46 "The worst mistake . . . drove it out." Rosen, *Minor Works*, 180, 183.
p. 46 "Such grave evils . . . own hands." Rosen, *Minor Works*, 184.
p. 46 "before . . . country." Rosen, *Minor Works*, 184.

p. 47 "For . . . enough." Rosen, *Minor Works*, 185.
p. 49 "2 and . . . midnight." *De rev*, IV, 5 (Wallis, 187).
p. 49 "In this area . . . error." *De rev*, IV, 3 (Rosen, 177).
p. 49 "In expounding . . . the Earth." *De rev*, IV, 1 (Rosen, 173).

CHAPTER 5

p. 52 "Faultfinding . . . poet." Rosen, *Treatises*, 93. Rosen first translated Copernicus's *Letter Against Werner* in 1939, from copies then available, and revised his translation in 1985 after examining several more copies that had come to light.
p. 53 "I therefore see . . . what is mine." Rosen, *Minor Works*, 312. This letter, translated from the original Latin by Edward Rosen, commands attention primarily for having survived the centuries, with its signature—"Nic. Coppernic"—intact, in the University Library at Uppsala.
p. 53 "To the Reverend . . . Copernicus." Rosen, *Treatises*, 93.
p. 55 "Some time ago . . . Nuremberg." Rosen, *Treatises*, 93.
p. 55 "Had it been . . . pleasure" and "I may . . . effort." Rosen, *Treatises*, 93.
p. 55 "However . . . fawner." Rosen, *Treatises*, 93.
p. 55–56 "Perhaps my . . . this subject." Rosen, *Treatises*, 93–94.
p. 56 "In the . . . time." Rosen, *Treatises*, 94.
p. 56 "We must . . . skill." Rosen, *Treatises*, 99–100 (slightly modified).
p. 58 "a second . . . the first," and "childish blunder," Rosen, *Treatises*, 104.
p. 58 "What finally . . . further." Rosen, *Treatises*, 106.
p. 58 "While referring . . . from love." Kesten, 229.
p. 59 "Undeniably . . . the weeds." Kesten, 229.
p. 60 "What kind . . . to say." Rosen, *Minor Works*, 189.
p. 63 "I saw Venus . . . at Frauenburg." *De rev*, V, 23 (Rosen, 276).
p. 64 "Some years ago . . . Farewell." *De rev*, front matter (Rosen, xxi).

CHAPTER 6

p. 66 "From one . . . the baskets." Rosen, *Minor Works*, 281.
p. 67 "My noble lord . . . acceptable." Rosen, *Minor Works*, 320.

p. 70–71 "Your Most Revered Lordship . . . your commands." Rosen, *Minor Works*, 323–24.
p. 71 "Therefore . . . bodies." Rosen, *Minor Works*, 326.
p. 72 "With regard . . . acceptable." Rosen, *Minor Works*, 331.
p. 72 "even . . . stable" and "it is unbecoming . . . church." Rosen, *Minor Works*, 340.
p. 73 "My lord . . . all my faculties." Rosen, *Minor Works*, 332–333.
p. 73 "I have now done . . . warnings." Rosen, *Minor Works*, 334.
p. 74 "Care should . . . higher judge." Rosen, *Minor Works*, 336.
p. 75 "I am sending . . . one another." Rosen, *Minor Works*, 336–37.
p. 76 "Together . . . judgment." Rosen, *Minor Works*, 337.
p. 76 "I most humbly . . . and help." Rosen, *Minor Works*, 343.
p. 79 "shy by nature." "those arts . . . crowds," and "It is characteristic . . . such things." "Preface to Arithmetic" in Melanchthon, 90–91.
p. 79 "Astrology . . . the planets." *Table Talk* DCCXCIX (quoted in Kraai, 12, n. 28). Chess Grand Master Jesse Kraai wrote his doctoral dissertation for the University of Heidelberg about Rheticus's efforts to establish a new astrology founded on a firm astronomical basis.

CHAPTER 7

Edward Rosen translated the Latin text of Rheticus's *First Account*, or *Narratio Prima*, into English in 1939. His rendering, quoted extensively in this chapter, remains the only English translation of the document.

p. 163 "It is . . . philosophy." Rosen, *Treatises*, 177–78 (except that "circle" is rendered as "sphere" in accordance with Swerdlow's "Pseudodixia," 122–23, n. 19).
p. 163 "Driven by . . . encourage me." Danielson, 139.
p. 164 "I had . . . weeks." Rosen, *Treatises*, 109.
p. 164 "To the illustrious . . . have studied." Rosen, *Treatises*, 109.
p. 165 "My teacher . . . method." Rosen, *Treatises*, 109–10.
p. 166 "We see . . . the world." Rosen, *Treatises*, 121–22.
p. 166 "A boundless . . . Amen." Rosen, *Treatises*, 131.
p. 167 "Indeed . . . alone." Rosen, *Treatises*, 136.

p. 167 "Hence you agree . . . phenomena." Rosen, *Treatises*, 140–41.

p. 167 "most nobly . . . golden chain." Rosen, *Treatises*, 165.

p. 167 "To offer . . . ," "Let me . . . ," and "But that you . . ." Rosen, *Treatises*, 115, 119, 128.

p. 167 "Most illustrious . . . goal." Rosen, *Treatises*, 186.

p. 168 "You might say . . . land." Rosen, *Treatises*, 190.

p. 168 "fisheries . . . ," "the illustrious . . . ," and "eloquent and wise . . ." Rosen, *Treatises*, 189–90.

p. 169 "At his . . . assignations." Rosen, *Treatises*, 371.

p. 169 "His Reverence . . . do so." Rosen, *Treatises*, 192.

p. 169–70 "Since my teacher . . . senses." Rosen, *Treatises*, 192.

p. 170 "compose . . . proofs." Rosen, *Treatises*, 192.

p. 170 "Then His Reverence . . . relied." Rosen, *Treatises*, 193.

p. 170 "By these . . . the world." Rosen, *Treatises*, 195.

p. 170–71 "When he . . . of office." Rosen, *Treatises*, 195–96.

p. 172 "from . . . Lutheranism," Rosen, *Scientific Revolution*, 161.

p. 172 "that Your . . . protection." Danielson, 79, 211.

p. 173 "This and other . . . troubled mind." Rosen, *Treatises*, 121.

p. 173 "to the . . . ," "splendid . . . ," and "Although . . . pass." Swerdlow, "Annals," 273–74.

p. 173–74 "Almighty . . . better" and "without . . . age." Rosen, *Minor Works*, 344.

p. 174 "recalling . . . sufferer." Rosen, *Minor Works*, 345.

p. 174 "are not . . . motions." Rosen, *Minor Works*, 344–45.

p. 174 "The peripatetics . . . the author." Rosen, *De rev*, 335.

p. 175 "contrary," "a true . . . astronomy," and "imploring . . . friend." Danielson, 212.

p. 175 "a greater . . . my friend." Letter translated in Danielson, 212–13.

p. 175 "Urania . . . I do." Gemma letter of July 20, 1541, quoted in Danielson, 116–17.

p. 176 "I have . . . negligible." Rosen, *Minor Works*, 350.

p. 176 "These writings . . . undergo . . ." Rosen, *De rev*, 351.

p. 177 "Highborn prince . . . doubt." Rosen, *Scientific Revolution*, 181–82.

p. 177 "Upon my departure . . . as he." Rheticus Dedication, *On the Sides and Angles of Triangles*, quoted in Danielson, 95.

p. 177 "Heliopolitanus," Kraai, 4, 74, 105.
p. 178 "So it goes . . . the Earth." Rosen, *Scientific Revolution*, 183.
p. 178 "the Polish . . . Sun." Letter of October 16, 1541, to Mithobius, quoted in Danielson, 91.
p. 178 "I regret . . . published." Rheticus's dedication to Heinrich Widenauer, quoted in Danielson, 98; also quoted by Westman in Gingerich, *Nature of Scientific Discovery*, 410.

CHAPTER 8

p. 179 "I confess . . . very questions." *De rev*, I, Introduction (Rosen, 8).
p. 179–80 "I can . . . stage." *De rev*, Copernicus Dedication (Rosen, 3; Wallis, 4).
p. 180 "They exhorted . . . proofs." Dedication (Rosen, 3).
p. 180 "the structure . . . its parts." Dedication (Rosen, 4).
p. 180 "After long . . . philosophers" and "against . . . sense." Dedication (Rosen, 4). For Copernicus's ignorance of the heliocentric model of Aristarchus, see Gingerich, "Did Copernicus Owe . . . ?"
p. 181 "Therefore . . . imagine." Dedication (Rosen, 5).
p. 181 "In order . . . mathematics." Dedication (Rosen, 5; Wallis, 7).
p. 181–82 "Perhaps . . . censure it." Dedication (Rosen, 5).
p. 182 "Astronomy is . . . the work itself." Dedication (Rosen, 5).
p. 186 "I was shocked . . . emergency." Rosen, *Scientific Revolution*, 165–66.
p. 187 "You have . . . profit." Rosen, *De rev*, xix; Gingerich, *Book Nobody Read*, 20.
p. 187–88 "There have . . . enough." Osiander Preface in Rosen, *De rev*, xx.
p. 188 "And if . . . him" and "Therefore . . . Farewell." Osiander Preface in Rosen, *De rev*, xx.

CHAPTER 9

p. 189 "Anyone . . . produced . . ." Gingerich, *Census*, 285.
p. 189 "so maul . . . future." Rosen, *Treatises*, 405.
p. 190 "On my return . . . sorrow." Rosen, *De rev*, 339.
p. 190 "I have written . . . Preface." Rosen, *De rev*, 339.

p. 192 "I am not . . . zeal." Rosen, *De rev*, 339.
p. 192 "a man . . . to appear" and "not . . . scholars." *De rev*, Copernicus Dedication (Rosen, 3).
p. 192 "I explain . . . weak." Rosen, *De rev*, 339.
p. 192 "for if . . . deceased." Rosen, *De rev*, 340.
p. 193 "acerbities" and "omitted and sweetened," Rosen, *De Rev*, 340.
p. 194 "She . . . to you." Rosen, *Scientific Revolution*, 169.
p. 196 "He has halfway . . . refused" and "He lay ill . . . him alone." Danielson, 217–18.
p. 197 "stimulated . . . bodies" and "to . . . Leipzig." Danielson, 129.
p. 198 "whose hand . . . this world." Danielson, 139.
p. 198 "I have not . . . width." Danielson, 139.
p. 198 "What sort . . . and devotion." Danielson, 143.
p. 198 "a sudden . . . unchristian," "minor child," and "plied . . . sodomy." Danielson, 143–44.
p. 199 "I have . . . stars." Danielson, 162.
p. 200 "again . . . commentary." Danielson, 172.
p. 200 "We had . . . of day." Westman in Gingerich, *Nature of Scientific Discovery*; Danielson, 191.
p. 201 "Twice . . . drowning." Danielson, 193.
p. 201 "I excavated . . . wonderfully." Danielson, 199.

CHAPTER 10

p. 202 "I deem . . . contemplate it." Caspar, 384.
p. 202–3 "In truth . . . world." Ferguson, 47.
p. 207 "I consider . . . astronomy." *Astronomia Nova* (Donahue, 43; Ferguson, 98–99).
p. 207 "burning eagerness." Ferguson, 155.
p. 208 "Days and nights . . . wind." *Mysterium* (Caspar, 63; Ferguson, 192).
p. 209 "I build . . . world." *Epitome* (Wallis, 10).
p. 210 "I was . . . ridiculous me." *Astronomia nova* (Gingerich and Ann Brinkley, quoted in Gingerich, *Eye*, 320).
p. 210 "sacred frenzy." Gingerich, *Eye*, 407.
p. 210 "If you . . . much time." Gingerich, *Eye*, 357.

p. 210 "hesitating . . . machine." *Mysterium* (Giora Han in Kremer and Wlodarczyk, 208).
p. 211 "It is . . . Copernicus." Gingerich, *Eye*, 300.
p. 211 "the unexpected . . . natural causes." *Rudolfine Tables* (Ferguson, 346).
p. 211 "Thee, O Lord . . . to that." *Harmonies of the World* (Wallis, 240).

CHAPTER 11

p. 214 *"The constitution . . . the word."* Dedication in Galileo's *Dialogue* (Drake, 3–4).
p. 214 "Would it . . . effort." Ferguson, 206.
p. 217 "Summon men . . . at all." Rosen, *Scientific Revolution*, 189.
p. 217 "concerning . . . humanity" and "read the . . . class." *Astronomia nova* (Donahue, 19, 21). Science historian William H. Donahue, who translated the *New Astronomy* from the Latin, says that Kepler's arguments on the interpretation of Scripture became the most widely read of his writings, often reprinted in modern languages, and the only work by Kepler to appear in English before the 1870s.
p. 218 "I believe . . . completely." Galileo's *Letter to the Grand Duchess Cristina* (Drake, *Discoveries*, 183–84). In addition to the passages quoted here from the *Letter to the Grand Duchess Cristina*, independent scholar Stillman Drake translated all of Galileo's works, most of which were written and originally published in Italian, not Latin.
p. 218–19 "To ban . . . of Heaven." *Letter* (Drake, *Discoveries*, 196.)
p. 219 "Now let us . . . teach us astronomy." *Letter* (Drake, *Discoveries*, 211–12.)
p. 220 "If we . . . stop too." *Letter* (Drake, *Discoveries*, 212–13.)
p. 221 "the whole system . . . sacred text" and "in . . . heavens." *Letter* (Drake, *Discoveries*, 213–14).
p. 221 "Grave . . . retardation" and "if . . . it resides." *Letter* (Drake, *Discoveries*, 214–15).
p. 222 "The words . . . in motion." Quoted in Drake, *Discoveries*, 164.
p. 222 "the quiescence . . . erroneous in faith." Consultants' Report on Copernicanism, Finocchiaro, 146.
p. 223 "false . . . Scripture." Decree of the Index, Finocchiaro, 149.

CHAPTER 12

In 1854 Jan Baranawski, a director of the Warsaw Observatory, published Copernicus's complete works in Latin and Polish—the first full translation into a modern language.

p. 226 *"I have . . . give it up."* Gingerich, *Census*, xxvii. Gingerich gave running commentary on his search for all sixteenth-century copies of Copernicus's book in academic periodicals, such as *American Scholar* and the *Journal for the History of Astronomy*. He also wrote a complete popular account, called *The Book Nobody Read* (Walker, 2004).

p. 229 "It is very . . . passage." Gingerich, *Census*, xxii.

p. 230 "For is . . . the Sun?" Gingerich, *Census*, 78.

p. 230–31 "Does the . . . irregularity." Gingerich, *Census*, 79.

p. 235 "So vast . . . Almighty." *De rev* I, 10 (Rosen, 22).

BIBLIOGRAPHY

Adamczewsi, Jan. *The Towns of Copernicus*. Warsaw: Interpress, 1972.

Armitage, Angus. *Sun, Stand Thou Still*. London: Sigma, 1947.

Bainton, Roland H. *Here I Stand: A Life of Martin Luther*. New York: Abingdon-Cokesbury, 1950.

Banville, John. *Doctor Copernicus*. New York: Norton, 1976.

Bartusiak, Marcia, ed. *Archives of the Universe*. New York: Pantheon, 2004.

Biskup, Marian. *Regesta Copernicana*. Wroclaw: Ossolineum, 1973. *Studia Copernicana* 7, in Polish.

Biskup, Marian, and Jerzy Dobrzycki. *Copernicus: Scholar and Citizen*. Warsaw: Interpress, 1972.

Blackwell, Richard J. *Galileo, Bellarmine, and the Bible*. Notre Dame, IN: University of Notre Dame Press, 1991.

Brahe, Tycho. *Instruments of the Renewed Astronomy*. Translated and edited by Alena Hadravová, Petr Hadrava and Jole R. Shackelford. Prague: Academy of Sciences of the Czech Republic, 1996.

Bunson, Matthew. *The Pope Encyclopedia*. New York: Crown, 1995.

Caspar, Max. *Kepler*. Translated and edited by C. Doris Hellman. Mineola, NY: Dover, 1993.

Christianson, John Robert. *On Tycho's Island: Tycho Brahe and His Assistants, 1570–1601*. Cambridge: Cambridge University Press, 2000.

Copernicus, Nicolaus. *Complete Works*. Translation and commentary by Edward Rosen. Baltimore: Johns Hopkins University Press, 1978 (vol. 2, *On the Revolutions*) and 1985 (vol. 3, *Minor Works*).

———. *On the Revolutions of the Heavenly Spheres*. Translated by A. M. Duncan. New York: Barnes & Noble, 1976.

———. *On the Revolutions of the Heavenly Spheres*. Translated by Charles Glenn Wallis. Annapolis: St. John's Bookstore, 1939.

Danielson, Dennis. *The First Copernican: Georg Joachim Rheticus and the Rise of the Copernican Revolution*. New York: Walker, 2006.

Donahue, William H., trans. Johannes Kepler's *Astronomia nova* (1609). Santa Fe: Green Lion, 2004.

Drake, Stillman. *Discoveries and Opinions of Galileo*. Garden City, NY: Doubleday Anchor, 1957.

———. *Galileo's Dialogue Concerning the Two Chief World Systems*. Berkeley: University of California Press, 1953, 1962; 2nd revised edition, 1967.

Eisenstein, Elizabeth. *The Printing Press as an Agent of Change*. Cambridge: Cambridge University Press, 1980.

Espenak, Fred, and Jean Meeus. *Five Millennium Canon of Solar Eclipses*. Hanover, MD: NASA, 2006.

Evans, James. *The History and Practice of Ancient Astronomy*. New York: Oxford University Press, 1998.

Ferguson, Kitty. *Tycho and Kepler: The Unlikely Partnership That Forever Changed Our Understanding of the Heavens*. New York: Walker, 2002.

Finocchiaro, Maurice A. *The Galileo Affair: A Documentary History*. Berkeley: University of California Press, 1989.

Galilei, Galileo. *Dialogue Concerning the Two Chief World Systems*. Translated by Stillman Drake. Berkeley: University of California Press, 1953, 1962; 2nd revised edition, 1967.

Gassendi, Pierre. *The Life of Copernicus (1473–1543)*. Translated, edited, and annotated by Olivier Thill. Fairfax, VA: Xulon, 2002.

Gassowski, Jerzy, ed. *Christianization of the Baltic Region*. Pultusk, Poland: Baltic Research Center in Frombork, 2004.

———. *The Search for Nicolaus Copernicus's Tomb*. Pultusk, Poland: Leopold Konenberg Foundation—City Bank Warsaw, 2006.

Gingerich, Owen. *An Annotated Census of Copernicus' "De Revolutionibus" (Nuremberg, 1543 and Basel, 1566)*. Leiden: Brill, 2002.

———. *The Book Nobody Read: Chasing the Revolutions of Nicolaus Copernicus*. New York: Walker, 2004.

———. "Did Copernicus Owe a Debt to Aristarchus?" *Journal for the History of Astronomy* 16 (1985): 36–42.

———. *The Eye of Heaven: Ptolemy, Copernicus, Kepler*. New York: American Institute of Physics, 1993.

———, ed. *The Nature of Scientific Discovery: A Symposium Commemorating the 500th Anniversary of the Birth of Nicolaus Copernicus*. Washington: Smithsonian Institution Press, 1975.

Gingerich, Owen, and James MacLachlan. *Nicolaus Copernicus: Making the Earth a Planet*. New York: Oxford University Press, 2005.

Grafton, Anthony. *Cardano's Cosmos: The Worlds and Works of a Renaissance Astrologer*. Cambridge: Harvard University Press, 1999.

Grant, Edward. *Planets, Stars, and Orbs: The Medieval Cosmos, 1200–1687*. New York: Cambridge University Press, 1996.

Heath, Thomas. *Aristarchus of Samos: The Ancient Copernicus*. Oxford: Clarendon Press, 1913; Mineola, NY: Dover, 1981, 2004.

Heninger, S. K., Jr. *The Cosmographical Glass: Renaissance Diagrams of the Universe*. San Marino, CA: Huntington Library, 1977, 2004.

Hirshfeld, Alan W. *Parallax: The Race to Measure the Cosmos*. New York: Freeman, 2001.

Hooykaas, Reiner. *G. J. Rheticus' Treatise on Holy Scripture and the Motion of the Earth*. Amsterdam: North-Holland, 1984.

Kepler, Johannes. *Astronomia nova* (1609). Selected, translated, and annotated by William H. Donahue. Santa Fe: Green Lion, 2004.

———. *Epitome of Copernican Astronomy* and *Harmonies of the World*. Translated by Charles Glenn Wallis. Annapolis: St. John's Bookstore, 1939.

Kesten, Hermann. *Copernicus and His World*. Translated by E. B. Ashton and Norbert Guterman. New York: Roy, 1945.

Kloczowski, Jerzy. *A History of Polish Christianity*. Cambridge: Cambridge University Press, 2000.

Koestler, Arthur. *The Sleepwalkers*. New York: Macmillan, 1959.

Kraai, Jesse. "Rheticus' Heliocentric Providence: A Study Concerning the Astrology, Astronomy of the Sixteenth Century." Doctoral dissertation, University of Heidelberg, 2001.

Kremer, Richard L., and Jaroslaw Wlodarczyk, eds. *Johannes Kepler: From Tübingen to Zagan*. Warsaw: Polish Academy of Sciences, 2009. *Studia Copernicana* 42, in English.

Kuhn, Tomas S. *The Copernican Revolution*. Cambridge: Harvard University Press, 1957.

Lattis, James M. *Between Copernicus and Galileo: Christoph Clavius and the Collapse of Ptolemaic Cosmology*. Chicago: University of Chicago Press, 1994.

Lukowski, Jerzy, and Hubert Zawadzki. *A Concise History of Poland*. 2nd edition. Cambridge: Cambridge University Press, 2006.

Manchester, William. *A World Lit Only by Fire*. Boston: Little, Brown, 1992.

Melanchthon, Philip. *Orations on Philosophy and Education*. Edited by Sachiko Kusukawa. Translated by Christine F. Salazar. Cambridge: Cambridge University Press, 1999.

Newman, William R., and Anthony Grafton, eds. *Secrets of Nature: Astrology and Alchemy in Early Modern Europe*. Cambridge: MIT Press, 2001.

Prowe, Leopold. *Nicolaus Copernicus*. 2 volumes. Berlin: Weidmannsche, 1883; Breinigsville, PA: Nabu Public Domain Reprints, 2010.

Ptolemy, Claudius. *Almagest*. Translated and annotated by G. J. Toomer. Princeton: Princeton University Press, 1998.

Rheticus, Georg Joachim. *Narratio prima*. Gdansk, 1540. Linda Hall Library. Digital Collections. November 2010. http://lhldigital.lindahall.org/u?/astro_early,148.

———. *Narratio prima*. Translated by Edward Rosen, in *Three Copernican Treatises*. Revised 3rd edition. New York: Octagon, 1971.

Richards, E. G. *Mapping Time: The Calendar and its History*. Oxford: Oxford University Press, 1998.

Rosen, Edward. *Copernicus and His Successors*. Edited by Erna Hilfstein. London: Hambledon, 1995.

———. *Copernicus and the Scientific Revolution*. Malabar, FL: Krieger, 1984.

———, trans. *Complete Works of Nicolaus Copernicus*. Baltimore: Johns Hopkins University Press, 1978 (vol. 2, *On the Revolutions*) and 1985 (vol. 3, *Minor Works*).

———. *Three Copernican Treatises*. Revised 3rd edition. New York: Octagon, 1971.

Swerdlow, Noel. "Annals of Scientific Publishing: Johanes Petreius's Letter to Rheticus." *Isis* 83:2 (June 1992): 273–74.

———. "The Derivation and First Draft of Copernicus's Planetary Theory: A Translation of the *Commentariolus* with Commentary." *Proceedings of the American Philosophical Society* 117, no. 6 (December 1973): 423–512.

———. "Pseudodoxia Copernicana." *Archives Internationales d'Histoire des Sciences* 26, no. 98 (June 1976): 108–58.

Teresi, Dick. *Lost Discoveries: The Ancient Roots of Modern Science—from the Babylonians to the Maya*. New York: Simon & Schuster, 2002.

Toomer, G. J., trans. *Ptolemy's Almagest*. Princeton: Princeton University Press, 1998.

Van Helden, Albert. *Measuring the Universe: Cosmic Dimensions from Aristarchus to Halley*. Chicago: University of Chicago Press, 1985.

Walker, Christopher, ed. *Astronomy Before the Telescope*. London: British Museum, 1996.

Wallis, Charles Glenn, trans. Nicolaus Copernicus's *On the Revolutions of the Heavenly Spheres*. Annapolis: St. John's Bookstore, 1939.

———. Johannes Kepler's *Epitome of Copernican Astronomy* and *Harmonies of the World*. Annapolis: St. John's Bookstore, 1939.

Whitfield, Peter. *Astrology: A History*. New York: Abrams, 2001.

Wlodarczyk, Jaroslaw. "Solar Eclipse Observations in the Time of Copernicus: Tradition or Novelty?" *Journal for the History of Astronomy* 38 (2007): 351–64.

Zamoyski, Adam. *The Polish Way: A Thousand-Year History of the Poles and Their Culture*. New York: Hippocrene, 2004.

ILLUSTRATION CREDITS

p. iv: Illustration of the Copernicus world system from a 17th century atlas by Gerard Valk and Peter Schenk. Corbis image #PG6097

pp. xi and xii: Maps copyright © 2011 by Jeffrey L. Ward

p. 5: Bayerische Staatsbibliothek, Munich, Cod. Lat. #27003, folio 33

p. 21: Astronomy Library of the Vienna University Observatory

p. 47: medievalcoins.com

pp. 50, 191: Uppsala University Copernicus Collection

p. 54: Courtesy of Owen Gingerich

p. 61: Nicolaus Copernicus Thorunensis Archives

p. 62: Princes Czartoryski Museum, Polish National Museum in Krakow

p. 130: Thomas Suarez Rare Maps, Valley Stream, New York

p. 165: Musées Royaux des Beaux-Arts de Belgique

p. 181: Museo di Capodimonte, Naples

p. 185: New York Public Library

p. 204: Fredriksborg Slot

p. 209: Pixtal/Glow Images

p. 233: Captain Dariusz Zajdel, M.A., Central Forensic Laboratory of the Polish Police / AFP-Getty Images

p. 235: Robert Williams and the Hubble Deep Field Team Space Telescope Science Institute (STScI) and NASA

INDEX

A Note on the Author

DAVA SOBEL is the acclaimed author of the *New York Times* and international bestsellers *Longitude*, *Galileo's Daughter*, and *The Planets*, and the coauthor of *The Illustrated Longitude*. She lives in East Hampton, New York.